# HOW TO BUILD DRY-STACKED
# STONE WALLS

# HOW TO BUILD DRY-STACKED
# STONE WALLS

## John Shaw-Rimmington

FIREFLY BOOKS

# A FIREFLY BOOK

Published by Firefly Books Ltd. 2016

FIRST PRINTING

**Publisher Cataloging-in-Publication Data (U.S.)**

Names: Shaw-Rimmington, John, author.

Title: How to build dry-stacked stone walls : design and build walls, bridges and follies without mortar / John Shaw-Rimmington.

Description: Richmond Hill, Ontario, Canada : Firefly Books, 2016. | Includes bibliography and index. | Summary: "This guidebook teaches you how to build walls, shaping creative designs into reality without using any cement mix. The book covers everything you need to know to build a natural, long-lasting wall: foundation, packing and backfilling, slopes, bridge stones, coping, and weight-bearing stones in an arch, bridge or dome" — Provided by publisher.

Identifiers: ISBN 978-1-77085-709-4 (paperback)

Subjects: LCSH:  Dry stone walls – Design and construction.

Classification: LCC TH2249.S539 |DDC 693.1 – dc23

**Library and Archives Canada Cataloguing in Publication**

Shaw-Rimmington, John, author

How to build dry-stacked stone walls : design and build walls, bridges and follies without mortar / John Shaw-Rimmington.

Includes bibliographical references and index.

ISBN 978-1-77085-709-4 (paperback)

1. Dry stone walls—Design and construction.  I. Title.

TH2249.S53 2016          693'.1          C2016-902287-0

Published in the United States by
Firefly Books (U.S.) Inc.
P.O. Box 1338, Ellicott Station
Buffalo, New York 14205

Published in Canada by
Firefly Books Ltd.
50 Staples Avenue, Unit 1
Richmond Hill, Ontario L4B 0A7

Unless otherwise noted, all photographs are by the author.

Illustrations by Nick Craine

Printed in China

The publisher gratefully acknowledges the financial support for our publishing program by the Government of Canada through the Canada Book Fund as administered by the Department of Canadian Heritage.

**On the Bias**, opposite:
A stunning and surprisingly strong wall built in California with mica schist flagstones from Colorado.

# Dedication

N O ONE HAS been more helpful and encouraging, no one more thoroughly knowledgeable and keen to work with on this project than my wonderful wife, Mary. Somehow, she has always enabled me to be passionate about what I do without letting me lose perspective. I am indebted to Mary for keeping my spirits up in times of frustration. Her way of looking at things from the other side of the wall has kept me building straight and true. If this book is in any way successful in sparking your imagination or providing useful tips on building walls without mortar, it is because she has given me the time and confidence to write it. I'm ever thankful to have her standing alongside me, even as she somehow managed to take care of the myriad other day-to-day things that needed attention while this project was being completed.

The author stands beside a 5-foot-tall containment wall in Langdale, in England's Lake District, opposite.

# Contents

In *Rubble Helix*, opposite, two twisting dry stone pillars built with dolomitic limestone rubble are held together with 8-foot-long through stones.

# Working With Stone

I REMEMBER THAT MOMENTOUS day 30 years ago, when I was taught to use a simple hammer and chisel to trace a line around a particularly large, round hunk of granite fieldstone. I was encouraged to hammer the chisel over and over along that line. Occasionally, I turned the stone over and hammered a corresponding line on the other side. I was to keep hammering until, by some kind of masonry magic, the rock split apart to produce two flat faces.

I worked at that granite rock for what seemed like an hour. Nothing seemed to be happening. My hands wanted to give up. The mason who was teaching me came over now and then to say, "Keep at it. It should open up pretty soon."

Sure enough, after another 100 or so blows, the stone started making a different sound. I could see a faint crack beginning to appear. My masonry instructor returned to see how I was doing.

"Don't hit it any more in the places where you already see a hairline crack," he explained, "but keep chiseling where the stone doesn't appear to be splitting yet."

I kept at it until, amazingly enough, the stone split right down the middle, magnificently, like some stale, pre-sliced, over-sized hamburger bun. Two sparkling faces, never seen before by human eyes, stared back at me.

I instantly forgot how numb my fingers felt. This was the first successful collaboration between my mind and hands over stone. And yet it wasn't at all aggressive. For all the pounding and effort, there was no sense of any violation done to the stone. Instead, I felt I had released the inside of the stone into the world.

Next, I learned how to shape the halves into cube shapes by chiseling four straight sides around each round face. When I was done, my two chunky stones lay beside all the other shaped stones, waiting to be put into a stone foundation. I had split and squared my first small granite boulder. My hands were eager to get started on the next one.

Each time I look at old walls, I think about the hands that built them. Judging by the moss and lichen on the stones, the last time some of them were moved could have been over a hundred years ago. Each rock was placed by hands that are no longer with us, but there is a stony memorial to them in the silent stones in the wall. Antique furniture tools and dishes have value because we can imagine the people who made and once used them. So, too, with old stone walls. They evoke a respect for the people who built them, thereby leaving their timeless mark across the countryside.

HARDSCRABBLE IS A term for poor, rocky, often hilly tracts of land on which it is difficult, almost impossible, to grow anything. In the 1800s, when the areas that were considered "good for farming" had already been claimed, newly arrived settlers reluctantly began to appropriate hardscrabble. Beyond the initial clearing of the land to make it suitable for agriculture, landowners soon learned that this kind of farming required the constant gathering and removal of vast amounts of stones and small boulders that were heaved up through the shallow soil every spring by the frost.

There was little time to do much with these stones other than push them off into the hedgerows. They were considered to be good for nothing, but it was necessary to get them out of the way. Miles of stones lay along the borders of fields indicating property lines and forming crude enclosures for livestock. Occasionally, however, a hard-working farm family would commit to building something with the stones harvested on their property. That these people had time and energy left to build more formal dry stone walls on their land, given all the demands of running even a moderately productive farm, is tremendously impressive.

As we look at these walls now, they may seem crudely built, and given the challenges of hardscrabble farming, that's understandable. But more likely, they are simply showing the signs of normal deterioration that affect even well-built walls that have stood for a long time. Often, their haphazard look is either the result of abuse or a failure to provide the care owed these silent testaments to the hardworking people who came before us.

It is interesting that part of the word "hardscrabble" — scrabble — is now the name given to a board game in which random letters are used to form words. Recently, as I gathered rocks and began to build a new wall along a small section of farm property, it occurred to me how much building a dry stone wall has in common with playing Scrabble: Having to figure out, as soon as possible, how to use awkward letters like Q and J with other letters to form words. Needing to nestle in small words between the others. Setting up words to create a place for other words later. Always trying things out, creatively experimenting and using the material you have at hand in a completely surprising way. Holding back some of the better material to use later on. Tossing tiles back in the pile and picking fresh ones. And, occasionally, missing a turn and going off for a coffee, then coming back later and suddenly seeing how things can fit together.

THE ACTIVITY OF building models and then creating garden features with stone is not very different from what a sculptor does. The difference is that while a sculptor who works with stone normally removes material that he doesn't want, a stone mason making a dry-laid wall is adding material.

My father was a sculptor. He made bronzes from welded copper, and he also worked in polyester resin and wood. His pieces were usually figurative and almost always impressionistic. He was very successful as an artist, and it was because he was very good at what he did.

I didn't realize until recently how much in common the work I do and the approach I take with stone has with my father's vision and output as a sculptor.

STONES HAVE DEFINITE seasons. They wait every day, just like plants, until their time comes. In spring, they peek out from their wintry homes where they've been nestled in the ground and begin to warm themselves in the protracted rays of sunlight. They eagerly anticipate being picked up and put back into the wall or taken off to find a home in a new wall nearby.

I remember picking stones out of the fields every spring on the farm. We loaded them in the tractor and carried them off to the side of the field and simply dumped them. I'd ponder the variety of shapes and couldn't convince myself they were all useless. As each one passed through my hand, I'd ask myself, "What can I do with this one?"

I found out slowly that stones know what they are about. It took me even longer to find out what my relationship to them was. Even then, it took years of doing stone and brick restoration, years of working with them until I ended up becoming a waller.

It's not the money, not the honor, nor even the pride of workmanship; it's being close to such sane, loyal, undemanding, totally inspiring parcels of matter on a day-to-day basis that makes this job so satisfying.

It's about the genuineness of the relationship between the animate and the inanimate. Hours of picking stones off the land makes you peaceful, keeps you simple and ever-contemplating. Stones teach you to wait for change.

They teach you tolerance too. Just the pleasure of gathering rocks on your own and knowing you are going to be able to find ways for you and them to work together — this can be the best part of your day.

**Opposite: Stones seem to emerge from the fields to warm themselves in the early spring sunlight.**

# Chapter 1

# Using Proper Tools

**T**HERE ARE LOTS of how-to manuals. There are some very good ones out there, and there are a couple of really bad ones. I've a passion for collecting books about dry stone walling. When I see one or hear about one, I buy it. But even a good how-to manual can't replace proper hands-on experience.

A better handbook for the beginner may, in fact, be a book of poetry. The thoughts and ideas expressed in a good poem may quicken your sensitivity to what is actually going on when you lay stone upon stone. Aesthetics is hard to teach, and a how-to book will not help you very much in that department. Wordsworth, Emily Dickinson, Robert Frost or T. S. Eliot may come a lot closer.

**Sometimes, using a machine to move a stone is a necessary evil.**

## Taking the Gloves Off

IT ISN'T OFTEN necessary to wear gloves, unless it is unusually cold or the stones are wet (wet stones are just too abrasive to work with bare-handed). Otherwise, I prefer to work without gloves. My hands like to feel and understand the different shapes of the stones. Walling is an engagement with stone, and, though it's not confrontational, it is worthy of all the attention that the phrase "time to take the gloves off" brings to mind. It isn't hand-to-hand combat but rather hand-to-stone contact. We pick up stones to identify their qualities and potential and to understand how we can fit them in the wall we're building.

Think about it. We don't put gloves on to pick up a child or greet a friend. We don't wear gloves to play a musical instrument, unless we're in a winter parade. Wearing gloves while building a dry stone wall is like steering a car by looking at the GPS.

There is an expression for what happens when there is too much going on around you: "I can't hear myself think." When I'm forced to work with gloves, my hands have an expression that roughly translates as, "We can't hear ourselves feeling."

Walling hands can't hear when their extremities are muffled in clumsy work gloves. They can't feel or see the heart of the wall. They can't tell how large the gaps are or what the shapes are under and between the stones. These are all things I need to know while I'm working and thinking with my hands.

My hands held things all day today. They not only held rocks of all sizes and shapes, but they also held chisels and hammers of all sizes. Some hammers took two hands to hold. Everything I held required that my hands worked together. They never argued with each other or worked against each other. They augmented each other's capacity to split rocks, fit difficult stones together and get things done.

# Hands Versus Machine

Stones usually respond better to the human touch.

This intriguing contraption is an early rock-moving machine. It relies on a team of horses and a geared system to lift and move rocks. The wheel placement allows it to straddle big boulders. Once the rocks are hauled off the field, the "wagon" can straddle a dry stone wall as the rock is deposited on top.

*The machine has got to be accepted, but it is probably better to accept it rather as one accepts a drug — that is, grudgingly and suspiciously. Like a drug, the machine is useful, dangerous and habit-forming. The oftener one surrenders to it, the tighter its grip becomes.*

—George Orwell

Sometimes machines are useful. I am grateful for the occasional mechanical help, but most of the time I would prefer not to introduce what seems like an unnecessary and annoying number of complexities into the equation. My hands and body really are content to work with the other hired hands lifting and moving stone. Doing the work yourself, without machinery, is not idealistic or stupid. It is quieter, healthier, in some ways safer, often more efficient and definitely, on every level and at every stage of the process, less environmentally harmful. Among other things, there is always the incessantly annoying sound of those bleeping vehicles as they back up. It is distracting and not at all conducive to working creatively. The more strenuous elements are part of the satisfaction of doing the job. I can do just about anything a machine can do and have more fun doing it too.

# ⋮ Essential Tools

**1** A brick hammer is not a tool for breaking stones or forming new faces or surfaces. Use it instead to chip off a stone's bumps and sharp edges.

**2** With this small, flat pry bar, sometimes called a Wonder Bar, you can avoid pinching your fingers when moving larger stones into position on the top of your wall. Use the bar to take just enough weight off one side of a stone so you can slide it into place without moving the stones below.

**3** Use the ridged end of the blue-handled walling hammer to split stones as well as to peel away stratified layers of stone. It is also handy for pounding off rough surfaces. Hold the stone vertically, and with the hammer, strike the stone with glancing downward blows.

**4** A point is generally used to smooth the surface of a stone, which allows you to fit the stone more precisely. Use it in combination with a medium-weight lump hammer (pictured here with yellow tape around the handle), to take off high spots.

**5** A carbide-tip masonry chisel with a 2-inch blade and 1-inch shank is the stonemason's best-loved chisel. It stays sharper longer than a standard tempered-steel chisel, and it is chunky enough to shape large stones but still useful for more delicate shaping. It can edge, flatten and split stone and may be the only chisel you need on the job.

**6** One edge of this walling hammer is shaped more like an axe than a hammer. Use it to split stones along their grain. The larger, rounded end can be used to shape stones fairly accurately. Do not use it to pound a chisel.

**7** These two hammers are lump hammers of different weights and sizes. They are used to pound stone chisels but can also be used for breaking stones.

**8** A string line serves as a visual guide to help you keep walls level and straight.

**9** Use spray paint to define the line of the wall as you mark it out on the ground.

**10** Use a tape measure to make sure dimensions are accurate.

**11** A big wrecking bar is very useful in prying large stones into place at the base of the wall.

**12** Your eyes can sometimes play tricks on you: A level keeps things level.

**13** Gloves are helpful when it is cold or wet.

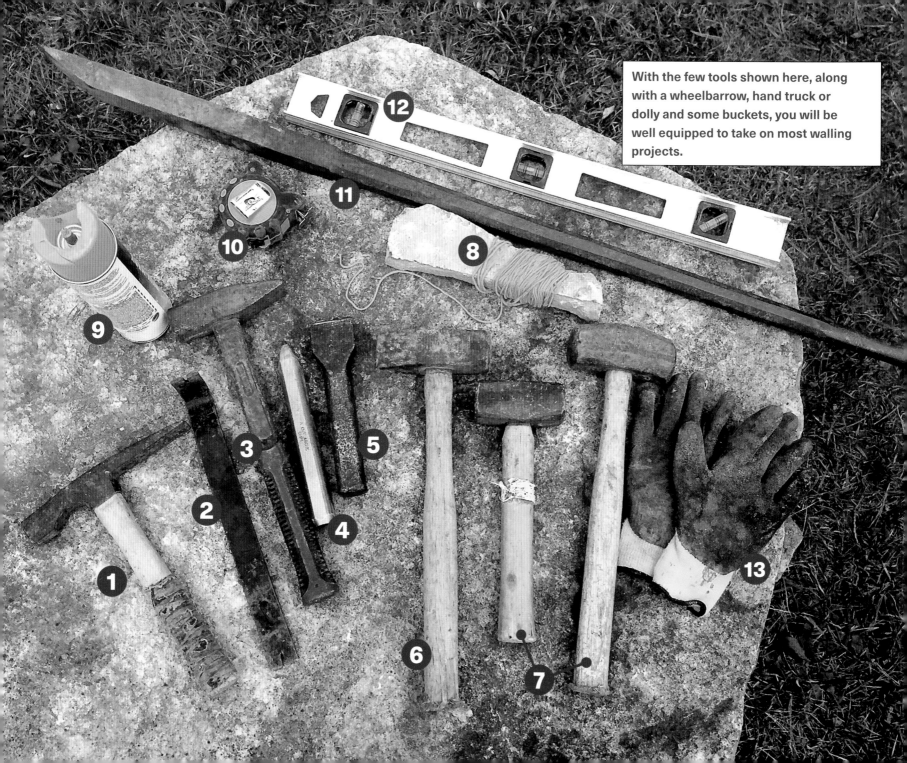

With the few tools shown here, along with a wheelbarrow, hand truck or dolly and some buckets, you will be well equipped to take on most walling projects.

# The Handy Wheelbarrow

There are often large stones that need to be lifted and moved. A wheelbarrow is the solution. But I don't want to try lifting this stone into the wheelbarrow. That will probably strain my back, and I might crunch my fingers or pinch my thigh on the edge of the wheelbarrow as I try to lower the stone into it.

Instead, I put the wheelbarrow on its side and roll the big stone into it. I put one arm on the lower side of the wheelbarrow and the other on the top; then I brace myself and push. Without any superhuman effort on my part, the stone and the wheelbarrow come to the upright position, and the stone nestles nicely into the wheelbarrow. Now I can easily move the big stone wherever I want it along the wall.

WE TALK ABOUT the way a thing "holds." A temporary wedge under a stone holds until we find a better-shaped one to fill the gap properly. A wooden handle holds the head that is fitted tightly onto it. The batter frame holds the string lines taut while we sight down the wall. The whole wall holds together because we have maximized the friction between each stone we have placed. We have made use of stone's three great properties: tremendous weight, great compressive strength and relative immutability.

Hands add another important capacity for holding dry stone walls together: flexibility. This flexibility is what is transferred to the structure of the wall when we don't use mortar. The wall virtually comes alive. It can move. It can yield to the forces of nature without breaking and falling apart. A wall holds, for all these reasons, and we hope it holds for a long time. My grip gets tired after a while. My hands need to take breaks. But the wall goes on holding. A day of me constantly holding and letting go of things is translated into years and years of stones holding together like some metaphysical equation. I am pleased with my investment of energy, my holdings. Everything is held in place, because nothing is withheld.

## Walling Hammers

IF YOU COME to dry stone walling from a masonry background, as I do, you face the choice of whether to use only the walling hammer or to continue using the masonry hammer and chisel.

The advantage of the walling hammer is that it's

**A lump hammer can be used to pound a chisel and break a stone.**

just one tool. The fewer tools you use, the fewer you have to lose. The walling hammer is efficient. It's simple. It's fast.

The hammer and chisel, on the other hand, can take off lumps that are almost impossible to cut with even the most deftly swung hammer. The hammer and chisel are also always more accurate. The walling hammer is slightly more likely to wreck a stone. Granted, a good waller could still use the misshapen stone in the wall, but sometimes you need to make sure you don't ruin the last perfect stone, the one that will fit perfectly into that special spot in the wall.

Stone is a very hard material. The common brick hammer and lighter carving tools can break when used to shape hard granite and limestone, causing

Hammer and chisel

Walling hammer

Handset

pieces of metal to fly off from the thin metal tips of these hammers and chisels. The best thing is to buy good heavy-duty, thick-shank chisels (preferably with carbide tips) and heavy hammers for squaring up stones. The proper combination of hammer and chisel feature metals of similar strength in each tool. Good tools last much longer than lesser tools before they start burring over or become misshapen and need to be ground down or retempered.

## The Handset

THE HANDSET, OR pitching tool, as it is sometimes called, has a very flat, blunt edge. It is a great tool for taking off ledges of stone along the face of a rock, even deep into the meat of a stone, and it can be very useful in removing stone and squaring it up. A tracer, in contrast, is a similar-looking chisel with a sharper edge. You can be more precise with a tracer, but it is generally not used to take off a lot of stone or used like a pointed chisel. (I, however, use it that way all the time. But that's not the point.) Next to the point, (a pointed chisel), the handset is a the hand tool of choice among most stonemasons and wallers.

## The Bull Set

USING THE BULL set is a two-person operation — one person holds a special bull-nosed hammer on its edge, set in place along the targeted area of stone, and slightly angles the blade toward the outside of the stone. The other person then hits it with the heavy striking hammer. The bull set hammer is not meant to be swung at all. But I've seen people do just that, breaking both the rules and the stones at the same time.

These hand tools are used to change the shape of stones through a grip-and-swing action, which is what our hands do very well. After using these tools for a time, our hands grow increasingly comfortable with them.

By comparison, power tools, with their constant machine-gun impact and violent vibrations, tire our hands and arms much more quickly. Our hands are continuously subject to the jarring torquing motions of some stone power tools. We must also watch out for dangerous whirling blades, screaming drills, fumes and on-site dust storms that make it hard to see and breath.

## Handling Bars

WALLERS FIND ALL kinds of bars useful. I use several types of wrecking bars to move huge boulders into different positions so that we can split and shape them. Some are rounded rods with a pry part at either end, some are square stock steel with a

**The bull set: Sometimes two hammers are better than one.**

# ⋮ Big Stones

**U**se the heaviest boulder-type stones at the bottom of a wall. That way, you won't have to lift them, and you can bury any irregular bottom surfaces in the ground and orient the better faces outward. If you use the bigger stones first, you can determine the number of irregular contour problems you will have as you work along the wall with the next course of stones.

The biggest stones in your stockpile should give you an idea of the minimum width of your wall. If the wall is too narrow, the boulders will stick out the sides. If the wall is only a little bit wider than your biggest stones, then you will not have left yourself much space to work on the sides of these stones.

I generally plan on leaving at least a foot or more on one side of the largest stones and let that determine the width of my wall.

Remember, too, that you don't have to use every big stone you

come across. Some of them will be too big or so round that they are almost impossible use, no matter how much you'd like to believe you can work them into the wall.

**Three men use pry bars to wrangle a massive stone into position.**

flattened wedge on one end that curves gently to about 10 inches for different applications when you are moving and tipping stones.

It is possible to "walk" a large boulder by resting the pivot point of the bar farther ahead in the direction you wish to go. Then push down and rotate the bar backward so the stone moves forward. The fulcrum is now repositioned and the rock can slide again in the direction you want it to go. A little bit at a time. Often, you have to move to the other side of the stone and alternate positions or you end up going in circles. Two people using these long pry bars can move large stones forward toward the wall.

Despite all the tools available, our hands are the prime movers when working with stone. Their strength is augmented by the simple yet profound physics of the lever. This way of moving heavy objects is a very natural and intuitive extension of the hand's capabilities. If my hands could speak, they would say they really enjoy moving big stones, some

# ⋮ Lifting Big Stones

**1** The first and most primitive way of lifting involves using wooden boards or a pole system. Simply slide or flip large rocks up strong boards onto the top of the wall. Here, you see how to lift a huge lintel stone into place.

**2** These three photos demonstrate a second method of moving large stones. With a simple pulley from a rent-all store, we lifted all the vault stones onto the Cabane roof. It was quiet, efficient exercise and extremely satisfying to build the entire structure without relying on heavy machinery.

**3** The third method involves a hoist, which is useful if you are lifting very heavy stones. We had to set up a hoist to lift the large rung stones that connected these dry stone towers. This structure was a very unusual shape and required scaffolding, from which we suspended a beam with a hoist attached. Then we hoisted the long, flat, heavy stones up near the middle of the paired spiraling dry stone walls.

**4** We often have projects that require us to stand in the bucket of a front-end loader filled with choice stones to finish the top section of a tall structure.

**5** Finally, if nothing else works, you can lift really big stones with a backhoe.

# Breaking Rocks

Swing the sledgehammer in line with the direction you want the stone to break.

Correct point of impact, shown here.

The result of an incorrect point of impact.

This very large, 20-pound sledgehammer is used to break up extremely large stones. It can break huge chunks of granite and limestone and any other type of rock we come across in our day-to-day work building dry stone walls.

It does a good job, but as with smaller sledgehammers, it must be swung carefully. Accuracy, not just strength, is required to break a big rock the way you want it to break. If you split it right, a large rock yields two halves. These halves usually have flat, straight faces that run at right angles to the length of the original stone. These split stones can then be laid structurally along a wall, with their faces showing out. Usually, at least with the quarried flattish limestone material we typically work with, a stone breaks better if it is hit off to the side rather than in the middle.

To get a crisp, long break, it is best to hit the rock near its edge (but not so near that you just chip a chunk off the side). When the stone is hit properly, the break spreads in one long, straight line across the stone. By contrast, if the rock is hit in the middle, cracks often radiate out from the point of impact, like a spiderweb, creating half a dozen short, less useful triangular pieces of stone.

far heavier than the weight of three men, with just a bar. They take great pleasure in not being relegated to the task of switching on and off some mechanized contraption that would make them redundant. If my hands could think, they would probably make sure I threw a couple of bars in the truck at the beginning of the day. They would come up with ways of using these bars to put big stones in the wall rather than letting me get lazy in my thinking and avoid including them in the first place.

## Measurement

TO FIND PERFECTLY sized stones as you are building a wall requires you to develop spatial memory. Identifying shapes and sizes and retaining that information as you are looking at a random pile of stones can be difficult. Perhaps the last time most of us practiced trying to find and fit differently shaped objects into various-sized openings was back in kindergarten, when we played with colored wooden educational toys.

You can use a tape measure to measure the particular stone you're looking for, but it's more efficient to measure the stone with your hands. Is it a hand width wide? That's about 9 inches, and it is referred to as a span. Perhaps the stone you want is two fists in length? Maybe an elbow plus an outstretched hand? Our hands are handy, always available and accurate. A tape measure is not only easy to lose, but it can't match our hands' ability to multitask as they

Rather than always depending on a tape measure, develop an ability to estimate the size of the stone you need for a specific spot in the wall.

sort, measure, carry and place. Part of the satisfaction of walling is finding simpler and simpler ways to do the job. Leave it to the hands to come through again and again with the sensible answer.

## Containers

COMMON FIVE-GALLON PLASTIC buckets are absolutely indispensable. Stacks of empty buckets can be found behind every commercial establishment, from small and large restaurants to industrial complexes, all of them just waiting to be picked up as garbage. For dry stone wallers, the backs of these buildings are prosperous hunting grounds for the treasured containers used to store and transport the smaller stones known as hearting material.

# Splitting Stone

The term "cleavage," introduced into geology in 1816, means the "action of splitting (rocks or gems) along natural fissures."

How cleanly and easily a stone splits is determined by a mineral's molecular structure and can be described as perfect, good or poor.

There is something wonderful about splitting open a rock along a straight, smooth plane. The hammer and chisel are not weapons, they are the instruments of an artist, a sculptor.

## Splitting and Shaping Stone

HUMANS HAVE BEEN taking on the demanding task of changing the shape of rocks and stones for a long time. Workers in stone in early civilizations in almost every part of the globe discovered all sorts of clever ways of prying and removing chunks of rock from the ground and cutting slabs away from outcroppings of bedrock. These big rocks were then made smaller, shaped perfectly and even decorated with textures and intricate designs. There are a variety of theories about how the Incas and other ancient peoples accomplished this with such accuracy and on such a spectacular scale.

The methods employed by the humble dry stone wallers, however, are not wrapped in any great mystery. Nor are their ways a closely guarded secret. Stacking stones is not rocket science. Nor is the shaping, splitting or even the dressing of stones. Only a few tools are needed, along with a certain understanding of some basic methods.

I believe that shaping stone should be differentiated from splitting stone in that shaping involves breaking a stone across its narrower thickness. Splitting, on the other hand, involves opening a stone up along or sometimes across its grain.

I rarely split stones unless I come across a stone that is layered and has a fault or a fissure where it seems it would be easy to separate it into two matching halves. The twin stones with newly revealed surfaces are sometimes called shiners. Shiners are rarely used structurally, that is, with their faces showing out in a dry stone wall, because they are

generally too thin, and the bulk of the stone can't be embedded deeply enough into the wall.

Shaping is used most often in normal dry stone walling. It is what is going on in the photo above. This stone is too long, and it is the wrong shape to go into the abutment we're making for the bridge.

In order to use this stone, we have to shape it with a chisel and hammer. The chisel has a carbide tip blade and a 1-inch shank, which makes it easy to grip and sends the hammer's blow with more force down into the stone.

Draw a chalk line on the stone as a guide. Using a 3-pound lump hammer, grip the chisel firmly near its pointed end. The chisel blade should be positioned on the line you want the stone to break along. Do not rest the chisel on a high point in the stone, which would direct the full force of the blow to one

small (directionless) point. Instead, make sure it is touching the stone's surface on at least two points. That way, the force of the blow is spread along the length of the chisel width and directs the force of the impact along the line.

Enough hard whacks eventually tell the stone where it should break. If the stone is fairly flat, I sometimes put a straight piece of metal (T-bar or rail or even a pry bar) under the stone, parallel to the line, so that the blow from the hammer is sent directly down through the stone to the metal. If I want to make sure the stone breaks in the right place, I sometimes turn the stone over and chisel the other side too. Eventually, the stone breaks.

Always remember, though, that you can never make a stone bigger.

**To shape a stone, place the chisel along the chalk line, ensuring the tip touches the stone's surface on at least two points, above left. If the stone is flat, place a metal pry bar underneath the stone, parallel to the chalk line, above middle. The tools of the trade, above right.**

Use a propane torch to remove ugly saw marks from sawn stone.

A tripod makes easy work of lowering heavy stones from a truck.

## Flaming

WHEN YOU HAVE special stones or are doing special fits, sometimes you have to make a difficult decision. In this case, a limestone slab was one of the last long through stones I had available, but it needed some work. Should I risk shaping it and perhaps breaking it with the hammer? The stone didn't have any kind of face on it, and I needed a flat face showing flush on one side of the dry stone wall I was building. So yes, I got out the noisy, dusty, evil stone saw.

The problem with saws is that they leave horrible telltale saw marks on the stone. It's difficult to rough up, or dress, these sawn stones with a hammer and chisel to get rid of these marks, and yes, you do need to get rid of them. The solution is to use a propane torch. Like magic, the surface of most stones can be roughed up with the flame.

## The Tripod

TO MOVE VERY large stones, I construct a tripod that consists of three 4×4 cedar posts connected at the top with a three-way metal bracket which pivots and allows me to adjust the leg spread. It's possible to move a heavy stone by placing a strap or chain around the stone and through a pulley attached at the top of the tripod. Lift the stone with the hoist, just to the point where there is almost no weight on the ground. Then swing the stone a foot or so over, lower it, and move the legs one at a time directly over the stone. Repeat, until the stone is where you need it to be.

**Norman Haddow, who introduced me to this simple and versatile walling tool, uses a tramp pick to move a big boulder.**

## The Tramp Pick

THE TRAMP PICK is a remarkably simple adaption of the heavy-duty pinch-point bar. It has some wonderful advantages for moving big stones. The tramp pick can be pivoted at a greater angle than the pinch bar, so there is less need to get a purchase stone under it for leverage. There is a place for your foot on either side of the tramp pick, and the point is splayed so it can be used like a shovel to get under stones. The handle makes it easier to twist and lift, and the bar can be used effectively to "row" a stone into place by not just pivoting it upward but sliding it horizontally along the ground.

**A dolly is one of the most important tools for moving large stones both to and along the wall.**

## Dollies

CARTS, WAGONS AND hand dollies, or hand trucks, are extremely useful for moving large rocks. They are quiet and maneuverable and far less expensive than bobcats or forklifts. With an industry-grade tree dolly, two men can gently lever a large stone onto the dolly and move it with little difficulty at all.

# Using Glacial Boulders

**E**rratics — boulders transported long ago by glaciers from their original location and deposited in a new home — often lie unexposed and undetected for eons in the thick geological strata.

I believe a thoughtful, more artistic, Zen-like approach toward these inspiringly mysterious visitors from down under needs to be encouraged among developers, homeowners and landscapers.

Several years ago, Northumberland County, where I live, put out a request for design submissions for a sculpture installation at the front entrance to its newly completed headquarters. A dry stone sheepfold design of mine, incorporating large boulders, was awarded the pubic commission.

We needed a lot of local granite field-stones. Fortunately, a new housing development was under way, and a pile of native stones had been unearthed during the grading and digging, shown in photo, above left. The stones looked as if they had been herded together and were waiting to be shipped off somewhere. I was pleased to be able to give them a home close to where they had lived for who knows how many thousands of years.

The finished sculpture, above right, was very well-received, and it has stood up admirably, considering it is situated near the busy public concourse. It does leave some people wondering about its function. I figure it's good to wonder about the meaning of art and stone, and their function and purpose, in both landscapes and creative installations.

**Job site before cleanup ...**

**... and after cleanup.**

# Cleanup

IT'S IMPORTANT NOT to leave material around too long after a job is complete. Grass grows up, making smaller stones hard to find, and these shards and chips become a hazard when mowing.

In addition, a tidy site instantly improves the look of what you have built. It makes it seem like the new structure has always been there. Photos are often taken right after a wall or arch is built, and it's always annoying to see tools, buckets and junk in those photos. A wall should look well settled into its environment.

Because it's hard to judge just how many stones you will need for a project, a lot of material is often left over. Sometimes, it can be stored and used on another project, but usually, it has to be taken away.

I use a 4-ton dump trailer to haul off my excess stones. I have also seen wall builders in the city use dumpsters. They fill them with wheelbarrows of left-over rubble rock and have them delivered back to their property or to the next job. It is a waste to have leftover stone dumped at a landfill site.

At the same time, it's surprising how often the leftovers can be used on the job at hand when you have no alternative material.

After the cleanup, plants become a big part of the lasting appearance of a wall. Dry stone walls don't exist in a vacuum — they need plants, and they need time to fit into the setting. Mature plants and shrubs soften walls and create a context for the new structure. Take time to come back and take photos of your wall a year or two after it is built.

## The Proper Height for a Wall

THE STANDARD WAY of establishing the height of a horse is by measuring how many hands high it is. This way of measuring may not have been very accurate initially, but today, the measure for a hand has been standardized at 4 inches. To offer some perspective, an animal that stands under 14.2 hands is still considered a pony, while an animal over 16 hands is considered a horse.

In most cases, a dry stone wall that is below 4-feet (12 hands) tall just looks wrong. It would be considered a pony, rather than a horse. A tall wall is elegant; it seems to have a purpose other than to provide a place to sit. The taller a wall, the stronger a statement it makes. We should take the opportunity to build proper freestanding walls when we can. Even a retaining wall, built to hold back soil, needs to have some height to it in order for it to do its job.

**Surrounded by the natural beauty of stone, this quiet spot is a relaxing oasis of calm.**

# Generator Wall

**A** wall like this can easily be built in three straight days or over two weeks, if you prefer to work only a few hours a day. It is satisfying work that can be done between barbecuing steaks or a few rounds of catch with the kids. It uses about a two tons of random quarried limestone.

One of the best things about a dry stone wall is that you can build one to frame or hide parts of your property. After annoyingly having our power cut off for 10 days over Christmas a few years ago, we decided to put in a generator. The unit had to be installed near the house, where it stuck out like a sore thumb. I ended up building a tall, semi-circular wall around the ugly plastic box that shielded the generator, using leftover limestone from earlier jobs. As an added benefit, we've also noticed that the loud sound the generator makes when it is running has been significantly muffled.

Walls such as these are useful for hiding other infrastructural eyesores as well, from propane tanks and electric transformer boxes to concrete well pipes that protrude above the ground. The walls don't have to entirely box anything in; they simply have to draw the eye away from the object that is currently getting unwanted attention. If there is ever a need to repair our generator, the wall leaves room to work, and if the generator has to be moved for any reason, the stone wall can easily be taken apart and rebuilt.

# Chapter 2

# Stone Walls

THERE ISN'T A property that can't be improved with the addition of a dry stone wall. What's noteworthy, however, is that the wall doesn't have to be a retaining wall — it can be any number of other structures.

Stones can serve many purposes in addition to simply holding back soil. If you are building with stones and fitting them together using the dry stone method, there are countless applications. The more you work with stones, the more you discover what can be done with them, and more importantly, what stones can do.

But a dry (also known as dry-laid or dry-stacked) stone wall, built without mortar, doesn't have to "do" anything at all. It can just be what it is — a freestanding structure that merely looks beautiful, the way a tree, sculpture or garden is beautiful in its own right.

**A house and yard before landscaping.**

**The yard is transformed by the addition of a dry stone retaining wall.**

The walls I discuss in this book are not a kind of ornamentation. Veneered stonework, by contrast, is basically just for show. The function of stones in most veneers is to look pretty. There is little regard for their inherent structural value.

A well-built dry stone wall, on the other hand, has an inherently pleasing form. That is because the wall has a fundamental and cohesive function. The thick stones in it have been utilized structurally.

This functionality can be extended to include purpose, as the proper walling method is combined with good design to create a formal or rustic feeling on any property. The dry-laid elements can be whimsical installations or artistic expressions, or they may have more specific applications.

A wall can be placed in the garden to enhance it or partially built around it to give a feeling of enclosure. Even a small section of wall can create a sense of peace and permanence. A wall can inspire our imaginations, suggesting a different time or place.

# Retaining Walls

IF YOU WANT to build a retaining wall, there are lots of things to consider. However, the basic structural elements are the same as those found in a free-standing dry stone wall.

A dry-stacked stone retaining wall should be thick enough to act like a dam, a dam that holds back not water and moisture but the weight and pressure of the soil. Otherwise, it is more like a "retaining hill," in that the earth is holding up the stone wall and not the other way around.

Since you are building a proper wall with soil packed behind it, the retaining wall may have to be thicker or be built farther into the soil. How high the wall needs to be is another question. Is there enough room to build a wall thick enough to do the job? It may be more effective to build two or three tiers, since multiple short tiers are generally stronger than one tall one.

Proper drainage is also critical. A dry stone wall drains much better than one assembled with manufactured "natural" stone blocks that are glued or cemented together. These sealed blocks do not allow moisture to percolate through, so they need drainage pipes that run through, under or around the wall to allow the water to drain away. By contrast, the many gaps in a dry stone wall allow moisture to evaporate and water to trickle through the wall. In addition, dry stone wall terraces and retaining walls

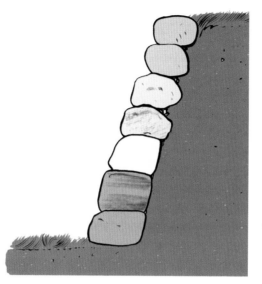
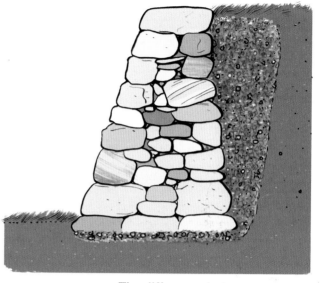

The difference between a retaining hill, above left, and a retaining wall, above right, is clearly seen in these illustrations.

end up being much simpler to build.

Use clear ¾-inch gravel under and behind the dry stone wall. Landscape fabric can also be employed as a barrier between the earth and the gravel to hold back any sediment buildup due to water seepage into the wall. In many cases, however, that is not necessary, and sometimes, it may even cause problems. If the fabric freezes or clogs, it can itself create a kind of water barrier. The more moisture trapped behind the landscape barrier, the more force is created by the frost or the hydraulic expansion in the soil. Clay soil swells with increased moisture much like the way frozen ground heaves, and both can have a destructive effect over time on a retaining wall.

In scenarios where the retained soil is largely composed of clay, laying drainage pipes behind a dry stone wall might still be necessary.

# Salt Spring Wall

**A**ndrew Currie is a 60-year-old Scottish dry stone wall builder and stonemason who lives on a horse farm on Salt Spring Island in British Columbia. He received much of his training working on the family farm in Argyle, Scotland, where he and his brother used to help their father make repairs to the dry stone walls that had been damaged, primarily by vehicles.

Andrew started this wall in 2000 and finished it before the end of the following year. They were three quarters through when the big earthquake hit Seattle. The wall held.

"There were 50 tons of cap stones at least on that wall, and it must have been at least 250 feet long and consumed over 400 tons of stone," said Andrew.

No material was purchased or brought to the island for the massive project; all the stone was found on the 120-acre property. The finished wall contains huge boulders (many of which had to be split), round granite rocks and lovely square-shaped chunks of sandstone.

"We used any stone, really. I call it using the 'stone of opportunity.' We used the tractor to move it all, and we used everything we found, and we needed a lot for hearting. It's a pain in the ass to have to make hearting stone, but it is the most important part of the wall."

# Sourcing Local Stone

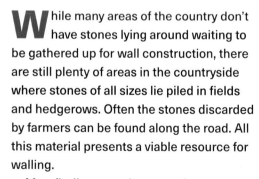

Collecting hearting.

Local quarries often have unpalleted rubble piles where wallers can handpick stone.

While many areas of the country don't have stones lying around waiting to be gathered up for wall construction, there are still plenty of areas in the countryside where stones of all sizes lie piled in fields and hedgerows. Often the stones discarded by farmers can be found along the road. All this material presents a viable resource for walling.

After finding out who owns the property, ask for permission to remove some of the stones. I have rarely had a landowner not consent to my request to collect their unwanted stones. More often than not, people are happy to have me take them away for free, but even if they are looking for compensation, $50 for a pickup truckload is a bargain for the best material to build dry-stacked stone walls.

## Quarries and Stone Suppliers

Local quarries are another good source of stone. Quarries often use huge mechanical crushers to turn the rock into various grades of gravel for construction purposes or to build roads. Perfectly good rocks for building walls are sent off instead to make gravel. I've often lamented that beautifully colored granite boulders and wonderfully shaped limestone chunks get consumed this way, without much thought for how valuable and useful they would be in dry stone wall building just as they are.

Fortunately, the quarries I deal with not only crush stone but sell random material "raw" by the truckload. Whenever I can, I have them deliver unsorted stone so that I get a much better variety of shapes and sizes, and I avoid a stacking cost. I often visit quarries to collect individual stones for specific projects such as art installations and footbridges.

## Freestanding Walls

WHEN YOU ARE considering building a dry stone wall, the first thing to determine is its placement relative to the property. A wall that runs parallel to the road is an obvious choice. The wall can act as a visual border, and it also dampens the sound of traffic. A wall can frame and protect a house and its surroundings better than a hedge or wooden fence. It can also be a very effective barrier against reckless drivers.

The telephone pole in the front yard of the house pictured here is missing a stay wire on the right: It was taken out by a driver who lost control of his car here before the wall was built. Luckily, there were no children playing in the yard at the time. The family is now very thankful to have not only a beautiful wall but a sturdy, functioning safety barrier too.

## Finding Stone

FOR US, IT is rewarding to collect stones during our trips back and forth from the cottage or during visits to friends in the country. The pleasure of spotting a well-shaped moss or lichen-covered rock, and the enthusiasm we feel carrying it home and imagining how it will fit in a wall is something very special.

The gathering of stones is not something that requires a special trip; you can stop the car any time you want. The task of gathering enough specimens to build a wall might sound like it would take forever, but the joy that comes from adding each handpicked

**A well-built dry-stacked stone wall compliments a house and property like a well-chosen frame completes a painting.**

stone to our pile offsets the magnitude of the task.

There is a strange thing about rocks and time. While onlookers may see the work as labor-intensive and a waste of time, it is in fact the very opposite: Gathering stones becomes a way of slowing down time. What could be more low-maintenance than a rock? A dry stone wall is simply a well-stacked pile of maintenance-free objects. Any section of wall we build, no matter how long it takes or how small it is, becomes a profound monument to our decision to step outside the demands of modern life with all of its hustle and bustle. When we build a wall, we enter a world that is calming and therapeutic. There is a healing benefit — a type of inner maintenance — free to anyone who takes the time to gather and stack stones.

# Bedford Street Wall

This wall on Bedford Street is a randomly coursed limestone retaining wall built to replace an original wall that had bordered the property for 50 years. It had eventually collapsed, most likely because it wasn't thick enough. In the rebuild, we reused the existing (brighter) stones and added limestone to beef up the dimensions of the wall.

My clients told me that one day, they watched as two men with a 4-foot spirit level got out of a truck and walked up to this dry stone wall in order to check whether the coursing was running exactly horizontal.

This slight optical illusion came about because the wall had been built on a hill that rose about 3 inches every foot. The courses do appear to be level, but they're actually not. They were built leaning slightly downhill. Sometimes you have to go with your eye and what looks right.

By seeing what looks level and measuring what actually is level, we can then decide what looks most pleasing with the rest of the landscape. It may be a compromise.

There are times when a dry stone retaining wall has an appeal similar to that of a natural band of exposed bedrock — which, of course, is never absolutely level.

# Step I: Arranging the Stones

**1** Locate and gather many more stones than you think you will need to build your wall.

If you are buying stones and having them delivered, figure on 1 ton of stone per 2 linear feet of standard freestanding wall, 4 feet high. That will give you enough choice as you build and leave you some left over.

If you are gathering stone, collect what you think are bad stones as well as good stones so you don't have to break the good ones to make rubble filler.

Always have enough selection available at the site to be able to work for at least the whole day. Remember, you need to have enough material to be selective as you work, since it's unlikely you'll be able to use every stone.

**2** At the site, move your material by wheelbarrow or hand truck and separate the stones by size. Place them flat on the ground and spread them out as much as possible so that you can see them. However, if the ground is likely to freeze, you may want to leave them in piles so that only the bottom stones are likely to freeze to the ground.

**3** When unloading stones, never throw stones onto other stones. They might break, and you can never make stones bigger again. If you must throw stones (off a truck for example), use the flat, more breakable ones tipped upright. If you are dumping stones on a driveway from a truck, first put down plywood and sand to soften the impact and cause less breakage and less damage to the driveway.

Separate and store candidates for through stones, the long stones used to tie the wall together.

Lay aside enough large, flat stones to use near the top of your wall and for coping.

Leave paths through the piles of stone so that you can easily walk away from your wall to find the right stones.

Leave ample room to walk along the sides of the wall as you build. This is also a safety precaution—you don't want to have to step over or on top of stones as you walk along your wall.

A spacious work site gives you room to make good decisions about fitting.

# Step 2: The Foundation

Establish a stable base for your stone wall with clear ¾-inch gravel.

A stable wall starts with creating a firm footing for foundation stones. Begin by digging a 6-inch-deep trench where the wall will stand. If the soil has already been disturbed, excavate to a depth where there is solid undisturbed ground.

In the trench, spread a clear, sharp ¾-inch gravel to a depth of 3 to 4 inches. The gravel provides drainage and a level support for the wall and should not consist of screenings, dust or any moisture-retaining material. It is serves as forgiving bedding for uneven stones and helps evenly disperse the weight of the larger stones used in the base.

Since any trench cut into the ground creates a place for water to collect, drainage is always an issue. A clear gravel base is less likely to compact or to create a dam that prevents water from flowing from one side of the wall to the other. If you do build a wall directly on the ground, remember that the weight of the finished wall may compact a sod or loamy surface, so be prepared for the wall to shrink a bit in height.

While it's also possible to build a wall in a trench without preparing a base, it is always better to place the foundation stones on gravel rather than directly on porous soil. Stones may shift when frost causes the soil to expand, but gravel offers air pockets in which ice can expand and then melt away.

No tamping is necessary, as the stones and whole wall do better nestled into the slightly yielding base material rather than teetering on top of an unyielding surface.

The base material does not necessarily have to be absolutely level. When you build a wall with random stones, as opposed to modular blocks, you can bring the wall to level by using a combination of varying thicknesses as you build.

# Step 3: The Foundation Course

**The foundation course should be laid level and below grade.**

The first course of stones is known as the foundation course. These stones should be placed, or bedded, into the gravel so that their faces are flush with the outside of the wall. Ideally, their heights relative to one another should be as level as possible to make building along and on top of them easier. This will take some doing, as large foundation stones, if they are not uniformly quarried stones, are often different heights and very different shapes.

With awkwardly shaped granite fieldstone, you can decide what the outside "face" will look like on the wall surface by determining what the worst side of your large stone is and then burying that side face down in the base gravel. This way, the gravel accommodates the awkward divots and rounded shapes. Make sure the bulk length of the stone is oriented so that it goes lengthways into the width of wall.

A foundation stone should be well supported by the gravel under it. Large cavities under big stones (especially flatter ones) can cause the stone to break or collapse into the vacant spaces.

If it is supported only at three or four points on the ground or by only a few smaller stones below, a course of foundation stones can eventually slump into the ground with the weight of the wall pushing down on them. That is why putting down a bed of gravel is the sensible step. The gravel provides even bedding underneath each stone.

If you are placing a very wide base stone, make sure there is still room to put a base stone flush with the outside of the wall on the other side. If it is too big, you may not be able to use it. Alternately, if you have a number of these wide stones, your wall may need to be widened in order to use them back to back.

Another solution is to bed one or two boulders well into the ground past the width of the wall and build the wall up and over them, conforming to the narrower typical dimensions of a dry stone wall. This is often the solution when bedrock or rocks that are too large to move are in the projected line of the wall.

# Step 4: String Lines and Batter Boards

The strings of the batter board provide the angle of batter and serve as a guide for aligning each stone face along the wall.

I typically use string lines when I am building a straight wall; to try to build something straight without them is a little risky. I occasionally untie them in order to move a number of stones to the other side of the wall or to access a tricky spot, but these lines have are really essential to building a straight wall.

But you can't just put string lines up and ignore them. Stones need to be lined up according to the level and plane that the string lines trace. You have to keep checking them.

There are two ways of putting up lines, both of which are useful. The first is to put up one line on either side of the wall, at the level to which you are working. When one level is complete, raise the lines to the next level. This is a very reliable way of building walls with all the same course heights. It helps, too, if you are creating or repairing a wall with random courses. However, it does make it difficult if several people are working on the wall at the same time. Inevitably, someone inadvertently leans into or pulls back the string lines while someone else is trying to sight down them.

There's another solution to this problem, especially if you are building with more random courses or working with irregularly shaped material. Patrick McAfee, a terrific walling instructor, heritage masonry author and stone-and-lime-mortar expert from Ireland taught me this method.

With the McAfee method, you set up two pairs of lines well above the height at which everyone is working. That way you can sight down onto the two lines and maintain the plane and batter, or inward slope, of the wall. Each stone that you place should have its face aligned along the same plane defined by these lines. All the while, people can lean over the wall or move stones without disturbing your sighting. Closing one eye and taking care to fit each stone properly will keep all the faces of the stones flush along three dimensions.

It is still important to step back from time to time to check that the stones are all running level and horizontally along each section of random coursing and that you are still covering every joint.

Whatever method you choose, secure the string lines to the

# Buckhorn Wall

There is an abundance of surface limestone and granite rocks in the Buckhorn area of Ontario. On this property alone, there was so much stone to be found in the hedgerows, as well as from the excavation sites for the pool and house, that no extra stone had to be brought in. There were some areas on the property where attempts had been made in the past to build with this gnarly stone, so we have tried to emulate this style in the walls we have built.

The chunks of stone, which are predominately limestone, are impossible to shape. One hit with a hammer causes the stone to break in totally unpredictable ways. The chunks of granite are too hard and too time-consuming to try to shape, except where the tips need to be knocked off to fit. Our client asked us not to use any pink granite in the wall. This was odd as the pink granite is found naturally here, along the gneiss, schist and dolomitic rock deposits. We obliged her as much as possible, though we did not completely banish the pink granite.

Normally, we stand flat copestones vertically along the wall to cap the walls (a style known as vertical coping), but since none of the older walls in the area featured this configuration, we decided to lay the larger flat stones horizontally. They do a good job of securing the smaller stones along the top of the wall.

batter boards (upright stakes separated by spacers/adjustable brackets) that are set up at either end of the section of wall to be built. The boards provide a guide to the size, shape and batter of the wall.

Dry stone walls are generally built with what is called a "batter," or a receding slope, so that if there is any movement of the wall (from frost, for example), the wall is able to shift back into its upright position rather than fall away. Walls without batter can last a long time, but they generally start leaning over in one direction sooner than batter-tapered walls do. A battered wall looks sturdier too. The closer to a pyramid shape the wall is, the sturdier it is, but an overly slant-sided wall starts to look less pleasing. The proper batter for a wall depends on the material being used to build the wall. The flatter the stone material is, the less batter is needed. Having said that, a typical wall generally leans in anywhere from 1 to 2 inches per foot of height.

A simple batter frame can be made with pieces of 1×2 wood

The dimensions for a batter frame, which prescribes the shape of the wall, are 3 to 2 to 1. The height is three units, the base is two units and the width at the top (minus the copes) is one unit. Thus a wall that is 30 inches wide at the base would be 45 inches high and 15 inches wide near the top.

screwed together so that the frame can be disassembled and reconfigured later with other wall dimensions. The batter frame should be secured with braces, screwed to a skid or propped up to something solid. Metal stakes can be driven into the ground, and these work well too.

**Opposite: A team of workers sorts, shapes and arranges stone in the first course of a dry-stacked wall.**

# Step 5: The Second Course

Placing the second course of stones — the first course above the foundation stones — is often the most difficult part of the wall-building process. These stones cannot be dug into the ground as you did with the first row. They have to be placed on a course of stones that sometimes turns out to be more uneven than you'd realized when you first laid them down.

Try to keep things level, and use slightly tapered bedded stones to make up for any fluctuations in height in the foundation row below. In cases where neighboring foundation stones are definitely not level enough, you'll need to go back and re-bed them.

With all other courses except the first foundation row, you won't have the option of changing the heights of the surface on which you're building; it soon becomes obvious that any stones laid in the first course that are not level with each other are a lot harder to straddle.

In general, after your foundation stones are laid, it is best to use a collection of your largest stones to build the second course. Your third course should be made with your third largest selection of stones, and so on, for each successive row laid.

Avoid the temptation to start shaping stones right away. Don't get frustrated — this part of the wall is going to test your patience and enthusiasm for developing true wall-building skills.

In coursing, the trick is to keep each row of stones sitting along a perfectly level line. That means the face and base of the wall should not only look straight, but every row should be level too. Otherwise, any variation in the heights of stones (especially neighboring stones in the same course) will become noticeable and throw off the line of the wall. Even a new wall with unleveled stones looks like it has sagged and is in need of repair.

The general rule is that stones should never be put in at an angle. If they are, that section of wall is not as structurally sound as it should be. (The exception that proves the rule is the herringbone pattern, in which all the stones are purposely put in at an angle.)

A slanting or crooked stone in a wall of level stones always looks wrong. Practice helps you get a feel for what is level, as does looking up at the horizon. Using a level tool is the most accurate method, but if you rely on it all the time, it will ultimately slow you down. To get the best sense of how level the stones are, stand back a few steps and take a look. That way, you'll notice if a particular stone is at an angle or if one or two stones look wrong compared with the others.

If you see a stone that doesn't look right, that's the time to fix it. Your first impressions in these cases are almost always right. I'm not talking about color or a nice shape but how structurally level the stone looks and how well it's fitting in with the rest of the coursing.

A stone that looks out of place will always look out of place. If you leave it and keep building on

# Irish Wall

This stone wall is just one of many unusual examples of wall-building styles you can see on the Aran Islands off the west coast of Ireland. The gnarly stones are gathered right off the weather-beaten bedrock surface of each of the three islands and built into walls to create windbreaks. Sand and seaweed are brought up from the beach in order to try to create soil for pasture, and the walls stop the wind from blowing away these precious beginnings of soil. The walls have to stand great gusts from the storms that come out of the Atlantic. Even though the walls are often only one stone thick, they are strong enough to last a long time.

The walls could have been built in the conventional way, but experience has led the local inhabitants to brace and wedge their stones against each other in an array of patterns that might confound those of us who are used to seeing stones neatly laid in courses and carefully butted against each other. There is a clever dynamic to building in a way that at first looks like a very random disorganized style.

The trick is to wedge all your stones into slots and openings created by appropriately large upright stones placed at intervals along the wall. If they are all placed properly, the wall can only get stronger if anything shifts.

These walls are a lesson in humility and prudence. The walls do the job without showing off or expending any extra energy. All of the available stones are used just as they are, and remarkably, they all stay together.

Stones and pieces of firewood should each be stacked perpendicular to the line of the wall or woodpile.

second round of cementing. It also won't involve the extra time required to wash off the cement and wait for the problem stone and surrounding stonework to dry before you lay the stones again.

It is generally more efficient to pick the stones nearest to you and find a place for them in your work area rather than searching for a stone to fit each specific hole. Shaping stones should be kept to a minimum. Build up the inside height of the wall evenly with the outside. Don't build the sides up at the expense of any interior structural soundness. Fill the insides with hearting as you go. Do not fill the inside of the wall with dirt, sod or anything that could retain water or might decay. You can use broken pieces of concrete asphalt, paving bricks, hardened bags of concrete or "ugly" stones for hearting. Using gravel is not a good idea as it adds very little structural value. If you use small, broken pieces of stone, they should be large enough that you can't scoop them up with a shovel.

Wall building is labor-intensive,

top, ignoring the message your eyes and brain have sent, every time you come back to that wall, you'll wish you had fixed it. By straightening it then and there, or replacing the stone that isn't level, you'll make it easier to build on top of it as well.

Thankfully, this is a dry-stacked stone wall, and the stones are not all freshly cemented together. Fixing a problem doesn't involve a messy dismantling and a

but it isn't backbreaking. That's why it is always funny to hear people wonder how we manage to do what we do all day long. If you are careful and don't try to be superhuman, the process is enjoyable. Remember not to lift heavy stones with your back, but rather with your legs. Or not at all—heavy stones can be rolled into place or moved with a dolly.

People who stack wood know they shouldn't lay a piece of wood along the line of the pile, but rather lay it *into* the pile. This same technique can be applied to stacking stone.

Most stones have some sort of length, so place them neatly and properly *into* the wall, lengthways. Even mortared stonework would be more structural if masons used this technique instead of laying them like vertical paving.

Try as often as possible to position the bulk of the stone inside the wall, not along the outer surface. Running stones along the length of the wall is called tracing and is not structurally sound. Nor should you sit a stone so that its bulk is positioned higher than its width.

**This wall was built almost entirely with round stones.**

## Round Stone Coursing

LAYING ROUND STONES in courses is not much harder than laying square or flat stones. When done properly, it looks stunning. I think it's worth learning how to do.

People often assume they can't build dry stone walls with round stones, but that is not true. When I have no choice but to use small, chunky, somewhat rounded field-stones, I definitely have to think more with my hands than my head. Amazingly, these stones are not as difficult to fit as they at first seem. In any pile of stones, there are very few completely round, ball-shaped stones. Most are shaped more like eggs or potatoes and have some length to them. The rule is simple: every stone should be laid lengthways into the wall.

The basic pattern should be a kind of layered honeycomb, not a grid or jigsaw. Place stones snugly up against one another in horizontal courses, and make sure they fit well on two surfaces. Lean each stone against its neighbor, working right to left, to create a dynamic force along the length of the wall. As they lean against each other, the bulky mass of each individual stone can also be set "leaning back" into the wall, so that even the roundest stones are trapped and won't fall out.

Think of round stones as square stones without corners. They are not really round at all because they all have some sort of length, even if they look round. They can be placed so that they lean inward and want to slip into the middle of the wall, rather than fall out of it.

Also, round stones placed together in rows have a lovely quality: At the top of each pair of round stones, a cradle is formed, which makes it easier to nestle the next course of round stones. That can actually add strength to the wall, as the stones not only push down but also spread out and tighten up to each other along the wall. There is a kind of dynamic energy running the length of wall, which a wall made with flat stones doesn't seem to have. Building a wall with round stones usually makes it easier to avoid running joints too.

It's strange how deceptively square these round stones look when they are painstakingly laid in courses in this way. That's probably why sometimes people who don't know a lot about walls comment that working with such "flat stones" must be pretty easy. They obviously think the stones are flatter and squarer than they actually are.

A randomly patterned wall may look less formal but can be as structurally sound as a more traditional pattern. The most important strategy in both styles is to fit the stones so that they are well bonded. Setting stones in a wall in a more irregular pattern takes more thinking, but it is often the only way to build, especially if the stones are angular and too awkward and varied in shapes to be laid in courses.

The care taken to fit a more challenging stone material of this type pays off. A well-fitted wall that creatively employs awkwardly shaped stone material can look far more appealing than a wall built with uniform, modular "overly shaped" stones that are simply laid up against each other in a predictable pattern (like common bricks and blocks).

The technique of laying stones in irregular courses requires a good deal of experimentation and creativity. "Think with your hands," and force yourself to pick up problem stones and try to discover the advantage of their unusual shape. A wall built in a random style is a wall where every choice has been a creative one, not just a predictable one.

There are two ways of going about building in this random style.

The first is to pick up any likely stone in your pile of material and walk along the section of the wall you are building to discover where it might fit. This involves placing the stone in several spots and turning it every which way so that you've tried all possible fits. The stone must fit structurally, that is, it has to be laid into the wall without creating a running joint (one or more vertical joints that run from one course to the next).

The second way involves remembering the general size and shape of the space you need to fill and searching through your pile to find it. It may take a few tries to find the specific stone you want, so this is a slower method and requires more spatial memory. Spatial memory is something a lot of us stopped developing after kindergarten, when we were last asked to find the peg that fit into a similarly shaped hole.

These two methods work best in combination. Generally, though, it is better to pick the stones nearest where you are working and find a place for them, rather than to go looking for a stone to fit each specific hole. Again, shaping stones to fit should be kept to a minimum.

# Herringbone Wall

**W**e used a herringbone pattern on the Irish Ditch built during a workshop we taught in Brockport, New York. This type of wall is wider than most dry stone walls. The sides are built with like-sized stones, and the wall is filled with smaller, less useful stone material. Lastly, the wall is topped with a covering of at least a foot of top soil.

In regular horizontal-coursed walls, stones laid flat only have their own weight (and the weight of some stones above them) to limit any shifting. But stones laid diagonally (as in a herringbone pattern) are further held in place by the weight of all the stones leaning to the left and right of them. Stones in a herringbone wall are more connected, and their dependence on each other is quite structural.

As long as the ends are built with anchoring boulders, the wall is extremely sound and strong.

# Step 6: Through Stones

In order to tie this double wall together, through stones should be placed every 3 or 4 feet.

Through stones are long stones that fit across the width of a double wall. They are important because they tie both sides of the wall together. Lay them at a height of 12 to 18 inches along the wall at 3-foot intervals to bind the wall together. If the stone you have isn't long enough, lay two of your larger stones adjacent to each other, making sure they straddle the stones in the wall rather than continuing a joint up from the stones below.

Sometimes, through stones are so long that they extend beyond the wall on one side or the other. If you trim them so they're flush with the wall, there's a chance they may break and will no longer be suitable as through stones. Sometimes, through stones protrude only on one side of the wall, making the other side of the wall look tidier. In Scotland, the tidy side is known as the "Laird's side," as this was the side the lord of the manor could see from his house.

One reason these stones may have stuck out on traditional British walls was that the lord of the property wanted to make sure the hired wall builders were not skimping on them. More likely, however, the random stones long enough to serve as through stones were just too awkward to shape.

Nevertheless, while it's a good idea to add through stones at regular intervals along a dry-stacked stone wall for strength, I prefer not to have these high-profile thigh-smackers sticking out on either side.

# ⋮ Step 7: Hearting

Hearting stones should be nestled between but always slightly below the builder stones.

Many buckets of hearting stone go into a dry-laid stone wall.

Stone wall builders may disagree about some of the rules of walling, but there is one rule that is pretty much universal: a dry-stacked stone wall needs to be properly hearted. Filling the inside of the wall with small, sharply shaped pieces is the most important part of good dry stone wall construction.

In a well-built dry stone wall, almost a third of the stone used should be small hearting material, pieces no bigger than small lemons and ideally shaped like lemon wedges.

To find enough small stone to do the job properly is both labor- and time-intensive. It's a quick and easy task to fill a wheelbarrow with five medium-sized stones, but to try to fill that same wheelbarrow with very small wedge-shaped pieces might take over an hour.

I always make sure we have enough of the right type of hearting material on site. For most projects, we have one or two men breaking larger, oddly shaped rocks in order to create an adequate supply.

Until recently, Ontario quarries, almost without exception, didn't specifically make the 3-to-4-inch clear material ("railroad ballast" is the best description for what we want) needed for hearting. Instead, they supplied smaller gravel-like material or a 3-inch minus material, filled with tiny stones and dust. Neither is appropriate for hearting. Small round stones act like marbles between the bigger stones, causing the wall to eventually fall apart, while gravel slips down, leaving gaps and piling up at the base of the wall, where it pushes the stones outward.

Areas inside the walls should first be packed with the largest hearting stones. Lay them flat (if they have any flatness to them) and at right angles to the wall, if possible. Then add smaller and smaller stones until the entire cavity between builder stones on opposite sides of the wall is filled.

This approach stops the wall from slumping, imploding or

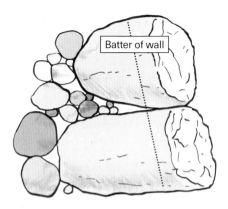

Round stones roll like marbles and so make poor hearting material.

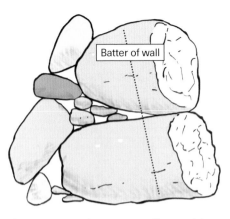

Placed to support the stones, these shims are ineffective and incorrectly hearted.

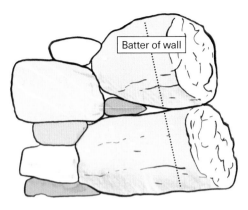

This section is shimmed with one stone and is properly hearted.

When packing the inside of the wall, carefully place each stone by hand rather than by a shovel- or bucketful.

being pushed in by external forces like people, animals, ice, snow and even strong winds.

Just as it's a mistake to carelessly cram as much as you can into a suitcase, it's a mistake to pack hearting stone sloppily. Take your time and place stones with matching shapes beside one another, putting smaller stones between bigger stones.

It's a hands-on thing. You're making a kind of tapestry. A wall builder should spend as much time thinking about the hearting as he does about the outsides of the wall. Just because no one else will see it doesn't mean it isn't structurally important.

This waller is breaking up larger stones for hearting material on site.

HOW TO BUILD DRY-STACKED STONE WALLS

# Indian River

We rebuilt this wall using all the original limestone, which must have come from the stony bottom of the Indian River, near Peterborough, Ontario. We were able to use several photos of the old wall as a reference.

Some retaining walls can be topped off with tight, thin stones laid vertically, as you see here. This is a typical way of finishing off the coping on a Cornish hedge. The copestones can then be covered in sod.

Sometimes, the garden soil stops flush with the vertical copes, and there are variations in which the copes sit higher than the level of the terrace as well. When the grass grows right over the stone, and you're standing at a distance, looking out over a property that has been terraced in this way, it's hard to see that there is a wall there at all. These invisible walls are sometimes called ha-has.

A ha-ha prevents farm livestock from encroaching on the more private parts of an estate. Landowners can still look out over their property without having fences or walls cluttering their view.

Jane Austen, notably in *Mansfield Park*, mentions the ha-ha when the character Fanny Price is forced to stay confined in the garden within its retaining walls.

# Building on a Grade

When you are building a wall on a hill, or "grade," with a rise ratio that is greater than 1 in 12 inches, you must first cut steps for the foundation into the earth. Then place the foundation stones, as shown in this drawing and the photo above. The height of the steps should be equal to the height of the foundation stones you are using. Courses above these foundation stones run horizontally along the wall and eventually join the copestones. Generally, the courses should be laid close to level.

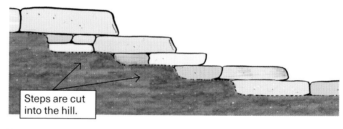

Steps are cut into the hill.

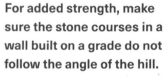

For added strength, make sure the stone courses in a wall built on a grade do not follow the angle of the hill.

Once the steps have been cut and the foundation stones are securely in place, each horizontal course of stones provides additional strength, right up through the wall to the copestones.

# Step 8: Bonding and Aesthetics

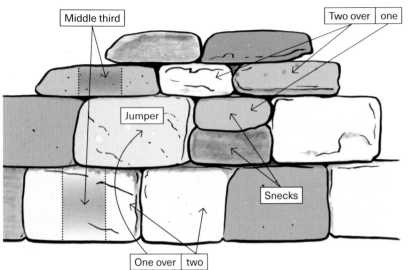

Middle third

Two over | one

Jumper

One over | two

Snecks

The "bonding" of the stones in a dry-stacked stone wall is of critical importance. The phrase that best describes how to do this is: one over two, two over one. Stones are laid in a wall so that they straddle *two* other stones. When two builder stones touch each other, that joint should be somewhere in the middle of the stone below them.

If you lay stones following this basic rule, you will avoid vertical joints (or running joints), in which a gap can run two or three courses up the wall. There is no connectivity at these places, and so they seriously weaken the structural integrity of the wall. Zipper joints, areas where gaps between stones create steep diagonal joints up the side of the wall, should be avoided too.

Every stone should be placed in the wall so that it touches neighboring stones in as many

This photo shows relatively good bonding work, in which the stones overlap one another. But there are some stones laid one-over-three, and that should be avoided. Can you see them?

places as possible. Friction and gravity must be maximized. Coursing stones along a wall helps maintain the likelihood that stones are bonded consistently, in the same way that bricks are laid. Crossing joints is not as easy with a random style of wall building. Where larger stones known as jumpers are used, which change the height of the course, two or three smaller stones, or snecks, should be stacked alongside to enable the stones to be straddled at the jumper height.

Try to create a consistent pattern over the entire length and height of the wall. This requires some discipline: Don't use up all of your good stones too early in the project. Know the sizes and shapes of the stones you have to choose from. Too many stones of the same size make a wall look uninteresting; too many varied sizes can produce a busy look. Arrive at a technique that uses the stones to address fitting and structural issues in the same way. Stones laid at an angle or piled in columns are less structural.

Round ones mixed with flat ones should be evenly distributed along the surface of the wall.

I have always maintained that building a dry-stacked stone wall doesn't have to be a trade-off between beauty and sturdiness and that both can be achieved with even a random assortment of stones. It is possible to erect something that is built like a tank, but is a tank really beautiful? A wall, unless it is built in a war zone, doesn't have to look like a tank-barrier either, just as a house surely doesn't have to look like a bunker.

Restraint is key for this to work. A wall built without restraint is a wall that wastes too many of the better-sized and better-shaped stones. Such a wall only demands more and more stones, which doesn't raise the standard so much as it does the cost. A wall will last a long time, but you may end up hating it because it isn't pleasing to look at. It's important to find the balance between beauty and practicality.

This section of a student-built wall in Queensville, Ontario, is a lesson in what not to do. It shows many examples of bad bonding as well as running joints, zipper joints, poor connectivity between stones and wavy rows. In addition, the plane of the wall wanders.

# Linnea's Wall

**S**tudents from Dry Stone Walling Across Canada (DSWAC) made this tidy-looking wall entirely out of very rounded glacial granite stones found on a property along the north shore of Lake Ontario near Grafton.

While the glaciers have not been so kind to the farmers and settlers in many parts of Ontario, they continue to provide an unending source of usable material for building beautiful dry-laid walls, as long as you're patient enough to learn how to build with them.

The rounded stones cannot really be shaped, but you shouldn't want to anyway. Rows and rows of round-faced stones laid in a wall are a pretty sight. The trick is to not have any of the stones set higher or lower than any of the others in the same row. Sometimes, just twisting the stone slightly can change its height so that a consistent level of accuracy is achieved along the rows. Each stone placed must also fully span the joint created by the two stones below it. The only flat stones in this wall are pieces of limestone brought in from a quarry to be used as through stones every couple of feet.

As you build up, each row diminishes in height and contains smaller and smaller stones of the same general size. There will be many stones you can't use, and no stones should be put in with their largest surface facing out. Once a combination of stones fit so that they don't fall apart or off the wall, most of the annoying gaps and holes between stones should basically be ignored. To fill these spaces with chips and small pebbles is to draw attention to spaces and does not accomplish anything structurally. In fact, if and when they fall out, the wall will be seriously compromised.

This wall continues around much of the front of the property, framing the old brick farmhouse beautifully.

# Step 9: Types of Tops

Triangular-shaped flat stones laid vertically work well as a top course for a wall. They can be shaped with a chisel or saw to look very formal and absolutely uniform. However, if there are enough triangle shapes in the stone pile, you may be able to do something like this with a lot less shaping.

## Vertical Coping

Placing stones vertically along the top of a wall is a technique called vertical coping. It is stronger than laying the same stones on their flats. Be sure to lay them so that their tops are all even and on the flat side. A wall with an even top looks tidy, even when the bottom line of the copes is uneven.

Vertical coping is preferred because stones that are laid on their ends exert more weight per square inch, which better holds down the flat outer stones at the top of the wall. A flat top can also look unfinished, as though there may be more added to the wall later. Other good reasons for using vertical coping are that it's harder for kids or animals to walk on vertically coped walls; stones stacked tight like books in a bookcase are more difficult to remove; and it adds an aesthetic appeal, like a capital to a column. Practically speaking, it also adds height to the wall quickly and

Flat flag coping is fine if you have enough large, flat stones to span the width of the wall.

allows you to use up the last of the awkward stones that couldn't be used in the main wall.

Before you begin, make sure you have enough stones of the right length to span the width of the top of the wall. Choose stones of similar heights. Put a larger, squarer cope at the ends of the wall. Lay a string line across the top of the copes at both ends, and place each cope against the end copes. Make sure none go above the line. Take stones out of the wall if necessary, so that taller copes fit level with the others at the top. Once the coping stones are all in place, wedge in small stones called pinning stones to secure them.

A Cornish copestone catwalk.

# Different Styles of Copes

An array of coping styles, clockwise from top left: ground cover, round stones, vertical coping and a mixture of shaped stones.

# Cumberland Walls

**S**ome time ago, we began experimenting with different turf-top applications to the dry stone walls on our property. The method the Scots use to finish off walls for which they don't have enough large flat stones is to dig up clumps of sod and lay them along the tops. The method involves placing a layer of sod upside down on the wall and then a layer on top of that, grassy side up. This wall covering works well in Scotland, where the soil is kept moist with the constant rain and drizzle of the Highlands climate.

We used rolls of sod on our walls and likewise placed the first roll upside down to allow the roots of the second layer (placed grass side up) to grow into the network of roots below. To ensure that soil doesn't trickle down through the wall, we put down a strip of landscape fabric before laying the sod.

Our walls looked good for that first year, but the grassy top suffered through the winter, partly because the layer of soil on the walls was so thin.

The next year, we experimented with thyme and sedum plantings. Both of these plants do better in dry, sunny conditions. We dug these plants into the sod top in the patches where the grass was not doing as well. Eventually, the sedum proved to be the best planting on the walls (dragon's blood is seen here in the foreground). The other patches of creeping thyme do well in wetter years. You can use allium, a pink flowering onion plant, too.

Some nurseries grow sedum mat material for use on green roofs, and these grow quite well on the tops of dry stone walls.

As you can see in this photo, the tops of our garden walls are thriving.

# Yoni Wall

The stones lying around your property are often the only ones available to you, but can you work with them?

These stones at first looked like real troublemakers. Not the kind you would choose to have on the job site. There was no getting around the fact that these particular combinations of stone shapes and sizes were very challenging. The wedge-shaped faces were not conducive to being laid flat and flush. Conventional coursing or random stacking were tricky, so we tried doing something different.

First, we laid out a footprint for two parallel curving walls, then laid the stones vertically, the way you would cope the top of a wall. We wanted to create the same protective and comforting feeling of enclosure along a length of pathway that you can experience walking along the narrow lanes in England's Lake District.

The two separate pairs of walls curved around behind a large redwood tree, where they opened up into a small enclosure.

# Chapter 3

# Stone Bridges

P EOPLE HAVE REMARKED on the natural look of the bridges I build. I guess the design has the simplicity of both form and function. A bridge made by hand, without machinery or power tools, usually has more appeal than one in which the material has been bullied and beaten into submission.

I am dedicated to the idea that bridges don't have to look pretentious. This means making them as efficiently and honestly as possible. I imagine many footbridges in Europe were built this same way.

A stonemason's skill is not determined by his stone-shaping ability alone. Nor is the calibre of his work a consequence of his having chosen the best material. Instead, his success can be measured by his ability to work with stones that a less-skilled waller might try to avoid. An important part of his talent is knowing when and where a challengingly irregular stone can be used to "do the job," looking just as it did when it came out of the quarry or was collected from the field.

# ⋮ The Arch Principle

One Friday afternoon a few summers back, I was in downtown Port Hope doing some banking. While I was in the bank, the friend I was with decided to do a little climbing. When I came out of the bank, I couldn't find him. He called down to me, and I looked up and took this photo of him.

It occurs to me that Evan's ability to hang there in space perfectly demonstrates the basic principle of the arch: As long as the sides don't spread (and Evan's abs hold out), he will not fall down.

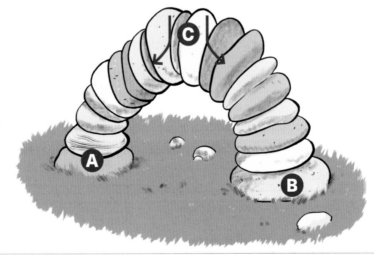

As long as A and B can't move or spread, C will stay in place.

A bridge built with restraint is less likely to look jarringly new and out of place, especially if the faces of the stones don't all have fresh breaks and chisel marks on them.

While it is important to fit every stone properly, building a bridge is not about impressing people with your stone-shaping skills. A bridge can be charmingly beautiful without being "showy."

Have you ever seen something new that looks so right, it feels like it has always been there? It takes a lot of invisible effort to make something appear effortless. I know I've done my best when I create a bridge in which the maximum structural integrity has been achieved with the minimum of fussing and bothering about shaping every single stone I've touched.

Anne Halliday's elegant dry stone bridge spans a stream at a golf resort near Glascow, Scotland.

## Abridgement

PERHAPS THOSE OF us who love dry-laid stonework and build with stone without mortar can better understand what it is about a dry stone bridge that makes it so profoundly attractive.

Many of us can certainly appreciate the rewarding sense of accomplishment Anne Halliday must have felt having recently completed her first bridge at the Westerwood Hotel & Golf Resort near Glasgow, Scotland.

Unlike thousand of bridges made with steel, wood or concrete, a properly built dry stone structure like Anne's beautiful arched bridge is not just a thing of beauty, it is also a delightful demonstration and resonating expression of what it is we mean when we say the word "bridge."

It is this expressive manifestation of the "dry stone bridge" alone that so mysteriously spans the

gap in our own imagination about what bridging is. We begin to understand not only why and how it stands, but what it is a bridge stands for.

Whle some of us will never get over the charm of a dry stone bridge whenever we have the pleasure coming across one, we recognize that it intrinsically embodies, in its unique structural design, all that is involved in the concept of "comprehending," literally "trying to get over" something. The stones are joined together in a magical pattern that reaches across the opening, and they rely solely on their own friction and weight to keep the whole arch suspended in air.

A bridge of this type epitomizes the artistry and craftsmanship involved and will always be more than the sum of the stones that constitute its parts.

Sadly, there are relatively few bridges here in Canada that are built of dry-laid stone. Not many can be found in Britain or Ireland or the rest of Europe either. Anne's project represents a growing movement, a resurgence in the interest in building permanently with stone and without mortar.

Fitting the stones together depends mostly on common sense and basic physics. Skilled artisans are helping demystify the concept of working with this plentiful structural material. All that is required is a dedication to building properly, the way people used to do in the past, long before manufactured products and highly specialized technologies came along.

## A New Stone Bridge

NOT LONG AFTER it was built, I was taken to see this very old-looking dry stone bridge. I am not that old. Nor in fact was the bridge. It had been built by master craftsman Norman Haddow and waller Dieter Schneider several months before. It was, however, the first dry stone bridge I had ever seen.

While visiting Scotland in 2003, I had an opportunity to meet Norman and travel with him to see examples of good-quality drystane dyking. He was eager to show me a bridge that had just been built from local boulders and a random selection of the mica schist stone found on a large estate located near the village of Butterstone in the Highlands.

We drove up some narrow roads and then walked quite a distance to get to what seemed like a very remote area. As we walked, Norman informed me that the design of the bridge was based on the bridge at Glenn Lion, which he had visited many times as a boy and had always fascinated him.

Norman had been commissioned to build a new bridge to replace a wooden one that spanned the burn (the Scottish word for stream or brook), where the client had proposed to his wife many years before. It was to be a very special project to commemorate the engagement and a successful marriage that had lasted five decades. At first, Norman was reluctant to take on the task, but after some persuading by the landowner, he agreed to take on the project. I'm glad he did as it was from that bridge that many other bridges arose.

When I first saw it, I had a bit of an epiphany.

**Opposite: This dry stone bridge built by Norman Haddow in the Scottish Highlands inspired the author, seen here, to build his own stone bridge back in Canada.**

**Digging a hole by hand for the bridge foundation is usually not that difficult and avoids all the mess and ruts that large machinery can cause.**

side of the burn. The boulders protruded upward, roughly to the height of the water during spring runoff. Next, a wooden form intended to temporarily take the weight of the bridge arch was carefully fixed in place by local carpenters. The supporting wedges could later be removed and the form slid out from under the structure. Once the two sides of the arch were built and joined at the top, the bridge would be self-supporting.

Perfectly flat, suitably thick, square-shaped stones work best in the construction of a challenging bridge like this one. Here, the valuable flatter stones, called voussoirs, were carefully laid in bands, basically upright with their best faces down across the wooden arch. Each row was leaned up against the previous one, ensuring that all of the rows gently leaned in a radial pattern up against the initial row of larger "springer" stones. That was repeated until the rows of stones almost met in the middle of the bridge.

Finally, across that remaining gap, a row of the best keystone-like stones were wedged in securely. All the stones that radiated over the form were then pinned with granite wedges of varying sizes. Pinning and wedging the structure thoroughly ensured that even the slightest wiggling of the larger stones in the arch would be eliminated.

The rows of stones at this point in the construction look quite jagged. A second and third layer of more horizontal stones was then built into and over top of the radiating rows of voussoirs, so that the arch grew thicker and heavier. This extra weight actually helped make the bridge even sturdier. As long as a

I understood for the first time that it was possible to span distances by carefully fitting random stones together in a proper arched configuration. After that day back in 2003, I became obsessed with the idea of some day building a bridge somewhere in Canada.

Norman's bridge hangs on four large boulders that were found on site and moved with heavy equipment close to opposite sides of the burn, after which they were manually barred into their final position. Other very large rocks were bedded together in the earth over these boulders, creating the two stone cribs that formed the abutments for the future bridge on either bridge has proper, solid built-up sides (abutments) that are unable to move, it is a principle of physics that the stone arch only becomes stronger as additional weight is added to it.

## Springdale Bridge

LESS THAN A year later, and with Norman's help, we organized the first DSWAC dry stone bridge building workshop in Canada. A handful of enthusiastic stone aficionados signed up for the course, and that October, we made bridge-building history at Port Hope, Ontario's first Canadian Dry Stone Wall Festival.

The property on which I had chosen to build the bridge was in the town my family and I had recently moved to. There was a good bed and breakfast off the main street on a beautiful four-acre property with a gentle swale going through a grassy clearing near a huge Victorian house. I proposed the idea of building a bridge on the property to the owner, whom I had never met before. He was not as surprised or as cautious as I had expected. Instead, he listened intently and was quite enthusiastic as I told him I had chosen his property for a very unusual stone project. When I inquired whether he would object to the idea of me constructing a traditional Scottish-style dry stone bridge somewhere on his land, at a cost to him of only 24 tons of stone, he replied "Am I an idiot?" I took that to mean he was all for it. As we got closer to the starting date, he grew more and more enthusiastic and eventually offered to serve lunches to everyone who would be working on the project.

Before the three-day public event, during which I hoped to complete the bridge, a fair amount of prep work had to be done. Enough suitable stone material had to be delivered from a stone quarry I had recently discovered about an hour away.

The foundation holes had to be dug and the abutments built up on both sides of the creek.

The 6-by-4-foot squares had to be dug in below the creekbed, which fortunately was not flowing during that uncomfortably hot week. Large limestone slabs had to be lowered into the holes and fitted in a dovetail fashion to create a kind of dry stone crib below grade that would support the bridge.

The two abutments had to be so deep that they would not move with the frost. My hope was that if they did move at all, they would each move the same amount and at the same time and in the same direction, so that the bridge would stay tight.

This bridge was to be an experiment. I needed to see if dry-laid stone was a suitable method for building small foot bridges in Canada. I had already built many smaller dry stone garden arches and realized that as long as the matching bases had the same size and depth of footprint in similarly undisturbed ground, very little movement would occur across the arches.

# Springdale Bridge

**W**e built the Springdale Bridge during the 2004 Northumberland Dry Stone Wall Festival.

We began by sorting, moving, shaping and rolling the stones into place. A crowd of onlookers gathered and watched throughout that first day as a bridge started to take shape — hardly a festival, but it was still a very festive time.

There were no machines, power tools, mixers, backhoes or bobcats — just the sound of chinking hammers and the church bells off in the distance. It was as though we had stepped back in time.

The bridge was completed and the form ready to be removed by 4 p.m. on Sunday afternoon.

The first person to go over the bridge was a friend of one of the builders. He crossed the turf-topped bridge in a large motorized wheelchair. After that, we all marched over the bridge and celebrated with a glass of Scotch.

# ⋮ Step I: The Foundation

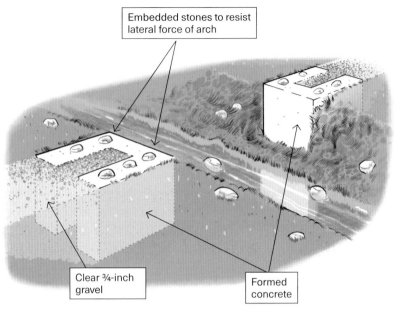

Embedded stones to resist lateral force of arch

Clear ¾-inch gravel

Formed concrete

**1** At the public bridge in Russell, Ontario, we dug 10-foot-wide, 4-foot-deep holes by machine and assembled U-shaped plywood forms in them. To make the abutments, we poured concrete into the forms; less was needed because of the clever U-shaped design. We then filled the area around the abutments with clear gravel, which was tapered back to support the tails of the bridge.

To introduce resistance to any sheering action across the top of the abutment's flat surface, we embedded stones in the fresh concrete.

**2** At the Hubbs Creek Bridge in Wellington, Ontario, we again used concrete forms but created 6-inch-high steps in combination with embedded rebar to act as resistance to the lateral push of the arch.

**3** At the Bruce Bridge, we dug precise holes in the hard subsoil that were 8 feet wide, 8 feet long and 8 feet deep. We then added clear, sharp ¾-inch gravel to within a foot of the surface. After compacting the gravel thoroughly, we built the bridge on top of this foundation. To avoid lateral movement in the bridge, we bedded the first large foundation stones into the gravel below grade. Several years later, the bridge shows no sign of movement or slumping.

**4** In the foundation for the Monarch Bridge near Cobourg, Ontario, we used large stones laid in combination with gravel fill.

# Step 2: Ribs and Lobsters

The wooden form (often called the "centering") that we place under the arch of a dry stone bridge until the stones are self-supporting looks a lot like an oversized lobster trap. The form assembled for the Hubbs Creek dry stone footbridge project near Wellington, Ontario, has a 12-foot span.

**1** The arc of this bridge follows the shape of a segmented Roman arch. To build it, we constructed a form in two sections from four sheets of ¾-inch plywood.

Having determined that the distance across the bottom of the segment was to be 12 feet (the width of the stream) and the height of the arc would be 3 feet on a flat surface, I fixed an axis and drew a circle with a radius where the line of a 12-foot span segments the circle at 3 feet along a perpendicular axis.

That gave me the height I needed for the form and thus the shape of the ribs for the arc of the bridge. I lay each of the four sheets down one at a time with one edge along the bottom segment and a corner aligned with my axis. Then I traced the half rib shape onto the plywood. The form's two halves were scabbed together with plywood squares. The curved "ribs" were equally spaced to create a form with a 12-foot span and a width of 6 feet.

**2** Before cutting the outside ribs, I drew radiating lines from the fixed axis while the plywood was flat. That way, I had guidelines to determine the angle at which each of the voissoirs should be oriented once the form was assembled and supported in place at the bridge site. It is always better to draw the lines at this stage than to try to guess the angles later.

**3** With a jigsaw or a skill saw, I cut a rib out. After I had cut one, I rotated the first rib around the plywood sheet. With proper placement, it's possible to get another full rib out of each sheet of plywood.

Plywood only comes in 8-foot lengths, so for the 12-foot span bridge, I made both halves 6 feet long at the base and 3 feet high. Building the form in two halves made it easier to transport to the bridge site.

The four ribs that made up one half of the form were spaced on 24-inch centers and secured with 2×4s and deck screws in five or six places. We screwed a couple of braces in place diagonally between the ribs to stop the form from leaning sideways with the weight of the stones. Several 6-foot-long 2×8s were screwed to the bottom of the form to give it stability. The other half of the form was made the same way.

At the site, the two halves were scabbed together and the form was then put into place. To make it lighter to carry, most of the 2×4s that served as the surface of the arch were added to the top of the form later.

# Step 3: A Good Platform

Some dry stone bridges can be built without our having to construct temporary bridges on either side from which to work. This bridge in Russell, Ontario, had very little rise, and the creek it spans was dry at the time. Our team was easily able to build the bridge and transport material across the creek.

HOW TO BUILD DRY-STACKED STONE WALLS

**1** The Bruce Bridge had a steady stream running under the form while it was being built. We were able to use the old wooden bridge as a temporary platform while we worked on the new stone bridge.

**2** At the Hubbs Creek Bridge in Prince Edward County, we created a working walkway by laying planks over the big barn beams that supported the bridge form.

**3** Because our 2×4s extended several inches past the form, we had to set up radiating lines for the alignment of all the voussoirs with strings rather than try to follow the lines drawn on the plywood.

# Cornish Hollow

My client asked me to build a bridge with a span of 12 feet — larger than that of the Springdale Bridge — so that she would have a way of getting her riding lawn mower over the creek.

Since the creek was in a protected area, I suspected she wouldn't be able to get permission from the conservation authorities. Three months later, however, she called to say she had got the permit. The structure had to be built in July, during a small window between spawning seasons for at-risk fish species.

I asked the local conservation officer, "How is it that you're allowing us to build a dry stone bridge over this creek?"

He said, "Many people apply to do all kinds of questionable things. Dams, diversions, ponds. We have to say no because of the damage that they would do to the water flow and the ecosystems. Now and then, in order to prevent landowners from just doing things without permission, we approve a few of what might be considered 'unusual projects.' Yours is one of those. The dry stone bridge you propose doesn't look like it will affect the stream while you build it, and the final structure will have a low environmental impact."

We set in place two foundations to support a 12-foot wooden form as well as temporary bridges. No supports were set in the water.

The bridge took 10 days to build. The heat was unbearable, so we used a canopy for shade — it was my first covered bridge.

# Step 4: Supporting the Form

The form used to support a dry stone bridge as it is being built can be held in place in a number of ways. Often, temporary beams or heavy planks are used. These need to be positioned so that they can be dropped down and the form lowered and eased out sideways once the bridge arch is completed and self-supporting.

**1** The crib of beams that support the form rest on big stones bedded in the creek. Wedges are slid between the top stones and the beams to bring the form to the correct height. These wedges are removed later when the dry stone arch is completed.

**2** The Springdale Bridge in Port Hope was supported by 4×4s laid on 12-inch blocks. After the bridge work was finished, the blocks were smashed and the form dropped.

**3** On a couple of bridges, I have used four pairs of these block wedges under the supporting beams.

**4** Plastic sheeting is sometimes placed between the wedges so that the wood doesn't bind when we spread them to lower the form-supporting beams.

# ⋮ Other Types of Arches

There are several different types of bridge arches besides the semi-circle arch and the segmented arch. One is the elliptical arch, top right. A superb example of an elliptical dry stone bridge was built in 2013 in Australia by my good friend, Gavin Rose. Visit the website where he describes the project at ttms.com.au/grampians-national-park.html

There is also the Gothic, or pointed, arch, bottom right. Gothic arches must be taller than are circular or segmented arches, and they are therefore less useful in crossing wide spans. In long bridges, they are used in combination. They typically require a higher foundation on both sides of the bridge as well. A flattened Gothic arch stone bridge was built over the Sudbury River in Aiken's Park near Hopkinton, Massachusetts.

**Elliptical arch**

**Gothic arch**

# Step 5: Springers

Springer

**1** Springers are special arch stones that are used in the first row of tapered stones from which the rest of the voussoirs "spring." Their bottoms are level, and their tops form a segmented arch that angles toward the center point of the circle below the form.

**2** To create a springer that is the right shape, we used a template to trace the proper angle on this large stone. Then, we sawed straight lines at the right depth and angle and chiseled away the remaining strips of material. In this photo, you can faintly see the saw marks left behind after the material was removed.

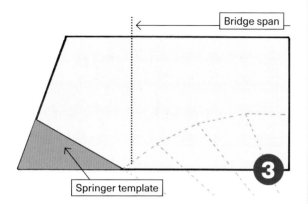

Bridge span

Springer template

**3**

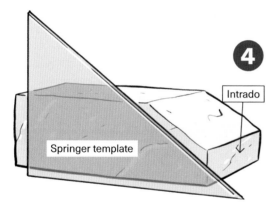

**4**

Intrado

Springer template

**3** This diagram illustrates the springer's angle and shape.

**4** Using the template, make sure the springer has some thickness at the nose rather than coming to a point. That way, it is stronger and less likely to break. This nose becomes part of the intrado, which is the term for the underside of the arch.

## Other Types of Arches (cont'd)

The catenary arch, top right, is very close to a parabolic curve, and there are many bridges that are good examples of this. These arches are usually stronger than segmented arched bridges.

The jack arch, or flat arch, middle right, is just that — an arch that has little or no curve. A bridge of this type requires maximum lateral resistance in the form of stone mass on either side of the span to prevent the arch from spreading.

A corbelled arch bridge, bottom right, relies predominantly on tensile strength rather than compression strength. Therefore, even if built with massive stones, it is a weaker arch.

**Caterary arch**

**Jack arch**

**Corbelled arch**

# Bruce Bridge

The Bruce Bridge was built between two ponds, to span a rushing creek that allows the upper pond to spill over into the lower one. As with most bridges, you can't really get a good look at it if you are just using it to get to the other side. This one is definitely more satisfying to see from a boat on the upper pond.

For this bridge, we used a 4-foot-deep, 8-by-8-foot gravel base—a daring diversion from previous foundations.

We also planted the top with all kinds of low-growing sedums. The idea was that they would do better than grass and could survive foot traffic. Ironically, they grow so beautifully that people are reluctant to walk on the bridge.

The round granite copes were a departure from the previous bridges as well. This was the first time we had mixed the two types of rocks found locally. They are kept in place by their sheer weight and because they are snugly butted up against each other. They not only act as an aesthetic visual border, but they are helpful in preventing a carelessly driven riding lawn mower from careening over the edge.

# Hubbs Creek Bridge

The Hubbs Creek Bridge was built as a landmark feature to attract visitors to a new winery located near Wellington in Ontario's Prince Edward County.

To start, we installed two large concrete piers below grade. We sourced old barn beams to hold the 12-foot span form in place. Since the local stone was too crumbly to use, we decided on a chocolate brown dolomitic limestone from Madoc, Ontario.

We had to visit the quarry to secure some special naturally shaped "springer" stones (they look like huge triangle slices of cake) and the two natural keystone pieces that were to be sandblasted with a decorative letter "K" prior to the building.

The build was set for August, when I was assured the creek would be dry and we could stand in it to set the form in. As it turned out, the creek was quite high and running swiftly.

Our goal was to use only a few tools, and the only machine was a tractor, which we used to haul stone around the property to reach the other side of the river.

On the last day, we added gravel soil and sod to the walking area.

# Step 6: Rows of Voussoirs

**1** On a dry stone bridge, the voussoirs are shaped as close to parallel-bedded trapezoids as possible and then laid in rows with their best face down and their longest side up. Here, the first voussoir is being pinned from the back with a piece of stone. The closer voussoirs can be fitted together at their faces (the exposed intrado side), the better.

**2** If the sides of the voussoirs don't fit well when they are butted together along the row, try other combinations until they do. Each odd-shaped voussoir should lock in somehow to the other. At the very least, it should overlap on one side to the one next to it.

Avoid using stones that are shaped more like triangles than squares. That is, avoid stones that are too narrow at their upper end when placed facedown on the form.

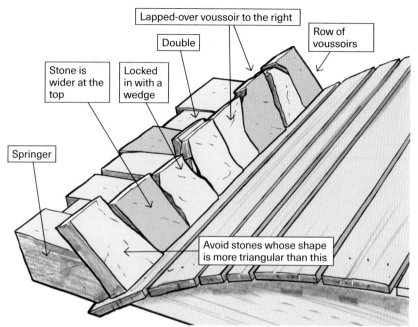

Lapped-over voussoir to the right

Double

Row of voussoirs

Stone is wider at the top

Locked in with a wedge

Springer

Avoid stones whose shape is more triangular than this

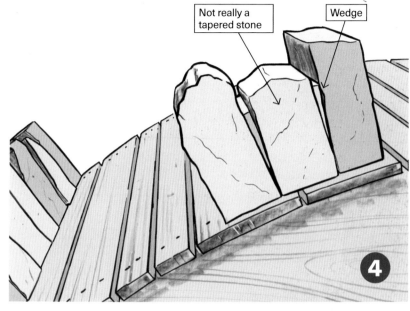

Not really a tapered stone

Wedge

**4**

**3** A row of voussoir stones should all be the same thickness so that the next stones can be laid across the joints. A pair of voussoirs can be laid together to make up the height of the other stones in the row.

As they are placed, pin these arch stones with thin wedge stones, so that each one fits snugly into the structure as it is being built. Avoid letting any wedges slip down and create pivot points between stones. The pins should be near the top and only used to allow the faces of the voussoirs to touch.

As you look across the bridge, the voussoirs should fan out slightly along the lines drawn on the form. These wedges should be pushed in just far enough that the voussoirs maintain their orientation. The final "shimming," which locks the stones in place permanently, is done when all the voussoirs in the entire arch have been set properly in place.

**4** Shaping each voussoir by sawing and chiseling every single one along its length to create the slight taper required for a row to fit in a radiating pattern and sit absolutely flush is an unnecessarily time-consuming job. Rather than shaping every voussoir, achieve the proper angle along the row by pinning each one, thereby effectively wedging it in place. With this method, it's possible to use many random-shaped stones that have little or no taper. If you pin enough of the stones in the bridge, it's possible to create a more informal, rustic look.

# ⋮ Step 7: Middle of the Arch

**1** As you can see, placing this middle voussoir was going to be a very tight fit. It needed to be snug but not so snug as to force the other stones out of alignment. Before the voussoir went in properly, some of the stone's thickness had to be chiseled away. If we'd forced it into the opening, it would have created a bulge somewhere farther along on the arch's curve, compromising the integrity of the whole structure.

These photos show some middle arch stones being fit between voussoirs that are already positioned across the form. On a segmented arch, the middle stones that complete the arch don't have to be special keystones, that is, they don't have to be bigger or different than the other voussoirs. As long as they are shaped to fit snugly into the arch, they will work.

**2** Once Evan felt confident that the chunky middle voissoir was going to fit at the MacDougall Park Bridge in Russell, Ontario, he hammered it into place with a heavy stone until the bottom touched the wooden form. The continuous curve created by the other arch stones was thus completed.

The other voussoirs required to fill in the middle space that extends across the form were all carefully fit and pounded in the same way. Some of these last interior voussoirs can be difficult to wedge in all the way to the bottom. If you cannot see whether the stones are flush with the intrados, the sound of stone hitting against the wood support tells you when they're down far enough.

**3** At the Hubbs Creek Bridge project, a special "keystone" had to be specially made. Prior to the arch being built, a suitable stone was chosen, and the letter K (for Karlo Estates) was sandblasted onto its face.

When the time came to fit this special keystone into the space we'd left between the other voussoirs, I was disappointed to see how loosely it fit. In order to correct it without having to take apart the other arch stones, we decided to cut out a portion of the form to allow the keystone to sit down lower and so fit more snugly between the voussoirs to the left and right. It was the perfect solution and ended up being a signature feature of the bridge.

As you can see, we were only able to "recess" the keystone because the 2×4s that made up the supporting curved surface of the form extended beyond the outer rib. This extension was added in order to build the Hubbs Creek Bridge 2 feet wider than the 6-foot-wide bridge form that had previously been designed.

# MacDougall Park Bridge

This dry stone footbridge was constructed in just two weeks in MacDougall Park in Russell, Ontario, near Ottawa. The original cedar footbridge was built around 1940 and did not survive the test of time.

The new bridge incorporates metal railings that are held in place by special steel brackets extending from the sides. These were embedded in the dry stonework as the bridge was being constructed.

The bridge spans 12 feet and used 24 tons of stone. Dry stone bridges have stood for hundreds of years in Europe, but to our knowledge, this is the first public dry stone bridge in Canada.

The bridge was built using primarily local materials and contractors. It was designed with a longer, gentler segmented Roman arch so as to be wheelchair accessible. A stately structure in harmony with its natural environments, the Macdougall Park Bridge is a frequently visited and an often photographed landmark.

# ⋮ Step 8: The Brackets

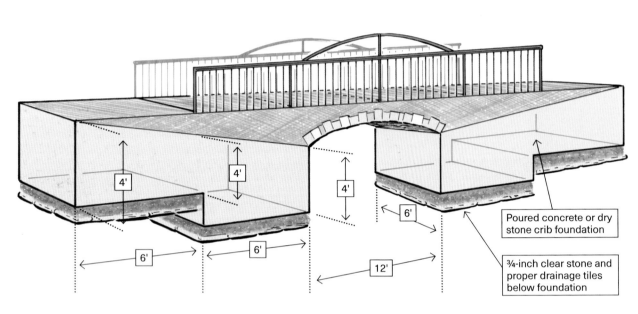

4' 4' 4'

6' 6'

6'

12'

Poured concrete or dry stone crib foundation

¾-inch clear stone and proper drainage tiles below foundation

The handrail strategy we employed at the Macdougall Park dry stone bridge in Russell, Ontario, was something Scott Cluett came up with to solve a common bridge issue: how to keep the width of the arch to a minimum and still have handrails.

It takes a great deal more time and work to build a bridge broad enough to accommodate wide copestone borders or structural dry-laid parapets. Securing the handrails beyond the 6-foot width of the bridge allowed all of the available walking surface on this bridge to be used.

After we reviewed the final design and dimensions, Scott, who is a part-time blacksmith, fabricated metal brackets from heavy steel with a thick rust-resistant coating. The

brackets span the width of the bridge and have sockets at each end. These are built into the bridge just above the voussoirs. Stones have been fitted around and then over them, with a final layer of pitched stone or thick flagstone paving laid over that, so that they are held securely. Metal posts were inserted into the square 4-inch sockets connected to the brackets, and handrails were then attached to the posts.

# Step 9: Copes and Strings

**1** Thin, flat unshaped stones were used to form a rustic copestone border on either side of the footpath at the Springdale Bridge. String lines were used to determine the general curve of the top of the copes over the bridge.

**2** Fatter, squarer preshaped stones were formally fit on the parapet of the Crown Bridge. String lines were used to keep the stones flush with the sides of the bridge.

String lines are useful at all times in bridge building to keep everything flush and plumb. Because the dynamics of the arch give a bridge enough strength, I do not batter the sides as I would with a narrower freestanding dry stone wall.

# Step 10: Taking Out the Form

Sometimes, the form is removed from underneath the arch before the bridge is entirely finished. In the case of the Hubbs Creek Bridge, one of my students had to leave before the two-week dry stone bridge workshop was over. At that point, all the voussoirs in the bridge were fitted and firmly locked in place with wedge stones and pins.

The arch was tight and almost all of the work on the sides of the bridge had been completed. Only the top needed to be finished. I knew the bridge would be able to hold itself up even without the addition of weight to the top, though that does give it even more strength.

My student was eager to see the form taken out before he had to go, so we smashed out the supporting cement blocks and the form dropped down a satisfying 3 or 4 inches. The bridge didn't move a fraction of an inch.

Next, we unscrewed the scabbed plywood that held the two halves of the form together, removed many of the loose 2×4s over the top of the form, and carried the right half away.

The second half gave us a bit of a battle as the form was still pinned at the northeast springer. With pry bars and some good whacks with a sledgehammer, the form came free, and four of us removed it. The arch looked strong and beautiful, and there was no concern that it might fall down.

# Monarch Bridge

**N**amed for the many butterflies that visited during construction, the Monarch Bridge is situated in a wooded valley on a beautiful property near Cobourg, Ontario. The bridge is entirely built of dry-laid stone brought in from a nearby quarry.

No mortar, metal or manufactured material whatsoever was used in this structure. Some of the limestone voussoirs, which were shaped and fitted over a temporary wooden form, were sawn and then chiseled into the proper shapes. The tail sections of the bridge were constructed with fieldstone.

The walking surface was pitched, or cobbled, with local rounded granite. On top of and along both sides of the bridge, we placed two rows of large blocky stones, flush with the outer edges of the bridge, and created an 8-inch border. After spreading a layer of very fine gravel, we embedded suitable 8-inch-tall stones inside the border to create a flat surface, much the way you do with cobblestone streets. We then brushed fine gravel sand between the cracks.

Wide enough for a golf cart, the bridge gives the property owners access to their acreage on the other side of the swampy creek that runs through their land. A wooden bridge would eventually rot, and it certainly wouldn't make the stunning impression that this bridge does.

# ⋮ Step II: Surface and Sides

**1** This is the way the back, or walking surface, of a bridge looks before a horizontal layer of flattish stones is added.

**2** After all the voussoirs are firmly set in place, the back of a bridge should be thoroughly pinned and then built over with stones. They don't have to be really good stones, but all of them need to be laid on their flat to within 6 inches of the finished height of the bridge's walking area. The parapet or border wall of copes are laid as you would standard vertical coping stones on a dry stone wall.

**3** A layer of gravel is added to the back of the bridge in wheelbarrow loads.

**4** Next, sheets of landscape fabric are laid over the gravel, followed by a layer of soil. Finally, rolls of sod are laid: the first layer is laid upside down; the next layer is placed grass side up.

A combination of level bedded quarried limestone and local granite fieldstone are used to build up over the arch. Coursing helps to keep things looking tidy.

## Building Over to the Sides of the Arch

MUCH OF THE visual appeal of a finished dry stone bridge is determined by the pattern of the building stones that form the approach to the arch and how well these stones join up with the individual voussoirs. Care should be taken to find shapes that match, which can be difficult because the tops of the radiating arch stones create acute angles. These triangular-shaped spaces are difficult to fill and can be an aesthetic problem as well as a structural problem. Too many small stones at the meeting line of the arch and the skewback (the sloping face where the end of the arch rests) risk looking busy, but they also increase the chance of the arch failing over time.

The stones in a bridge have to be laid into the structure much like those in a dry stone wall. They should have their longest sides oriented on their flat, perpendicular to the plane of the bridge.

Stones that adjoin the arch need to merge nicely,

HOW TO BUILD DRY-STACKED STONE WALLS

in an assortment of rows according to size away from the bridge. This makes it possible to choose stones of a similar width and prop them up together to see how they fit before we actually begin to use them in the construction.

To do this, find a flat area, and lay out enough plywood to accommodate the dimensions of the arch form, which is a rectangle the width of the bridge times the length of the arc over the span (this length will be slightly longer than the span across the bottom of the form).

Lay the stones face down, as if you were pitching cobblestones on a path, only upside down. Choosing all the voussoirs for one side of the bridge beforehand allows you to make better decisions about which stones work better together. Sorting and moving them around and trying out different configurations is far more convenient when they are lying flat on the plywood. Once the rows are placed on the bridge, you won't have the same opportunity.

The other benefit is that more people are able to work on the project at the same time without everyone crowded around waiting to put stones over the arch one at a time.

The final assembly of the arch also takes less time, since the voussoirs can be easily moved one by one and placed on the form. Each voussoir is spread with small wedges so that they all fan properly. When people are transferring the individual stones to the bridge, care should be taken not to mix up the rows or reverse the sequence of voussoirs left-to-right.

**In these photos, workers arrange voussoirs at a bridge we built near Montreal.**

but they should also be laid deep enough into the body of the bridge so as not to come loose. Again, if too little attention is given to the joint where the horizontal builder stones and the voussoirs meet, the structure will look crude and poorly settled.

## Preconfiguration

BY SPENDING TIME choosing the best stones for the purpose, rather than fitting them one at a time over the actual bridge form, we are able to pre-fit them

# Kay's Bridge

This bridge was built to replace the original wooden trail bridge at the Landon Bay Centre in Gananoque, Ontario, which was in poor condition. Kay and her husband Charlie were some of the original owners of the property, which they later donated to the town as a public park.

We decided on a completely dry stone arched footbridge with a 12-foot span, built on the side of a forested granite knoll over a craggy year-round stream. It was the perfect place to build a bridge since it would sit directly on bedrock outcroppings.

First, we cleared the area and then began chiseling, sawing and shaping the bedrock on either side of the creek. This was done to make steps and ledges to accommodate the sandstone springers from which the bridge voussoirs would "spring."

The covering of the bridge back was done in cobblestone "pitching," which turned out to be a very durable material for foot traffic.

As we built the bridge, the water level of the creek kept rising due to tremendous rains earlier in the week. We were worried the form would be swept away, but by the second day of the build, the water had leveled off just below the wooden arch supports. We completed the bridge in six days.

# Double Arch Bridge

Taking on the project of designing and organizing the construction of the Double Arch Bridge near Sainte-Marthe, Quebec, was a bit of a challenge. It was to be twice the size of anything I'd ever built. The bridge was constructed on a very picturesque property surrounded by acres of manicured lawns and vineyards.

Everything about the bridge was considerably over-engineered. The two 10-foot span arches required a middle pier foundation (known as a cut water) to support the double thrust. The foundations for the outer abutments and this middle pier were made of huge concrete blocks, the type used as dividers on highways and barriers in parking lots. These were laid in a crisscross pattern, 8 feet deep in the ground and up in 12×12 columns, up to within 10 inches below grade.

We needed a lot of stone material delivered (over 60 tons), and in the end decided on a local sandstone.

Approximately 25 people worked on the bridge, and the bulk of the dry stonework over two arch forms was successfully completed over three days.

In celebration of its completion, a Scots piper walked across the bridge before everyone involved (plus an audience of about 100 people) stood on the bridge for a photo. Then we drove a small front-end loader over the bridge just to prove how strong it was.

# Chapter 4

# Follies

OLD STONE WALLS often awaken in us a sense of wonder, accompanied by a curiosity about who built them and why. This is especially true when we accidentally come across what appears to be an ancient semi-ruined edifice.

It seems unusual that such architectural fragments and large stone relics should evoke feelings we have not felt anywhere else. Things that have remained perfectly intact don't usually produce the same delightful reactions.

With ruins, we have a unique sense of inclusiveness. To the degree that we pick up on the vibes of some lonely half-fallen-down, half-forgotten structure, we have an assurance that we have not been left out. There is a quiet knowing.

WHEN WE ENTER particular stone structures and are surrounded by their thick stone walls, we are somehow affected in a positive way by the sheer mass and density of this enduring natural material. Unconsciously, we intuit something so still, so enduring, so powerfully restrained, that there is a perceptible benefit to our being enveloped in this quiet energy.

While dry stone walling is a genuinely respectable trade with many serious applications, now and then it can branch off in whimsical directions. A number of years ago, I was asked to create an enclosed space in northern California with stacked stones and to make it look look like an abandoned stagecoach stop. The property did, in fact, have an old stagecoach road running through it. The owner wanted something on his stately acreage of tall redwoods to allude to the history of the place. We thought that could be best accomplished if we built a structure that would look like it had been there for a very long time.

We decided to build in a clearing off to the side of that abandoned, rather overgrown, stagecoach trail, in order to give the impression that travelers had stayed there many years ago.

No one had ever seen a dry stone stagecoach house (apparently, they were all made of wood) and there were certainly no pictures of such a thing on Google. That did not stop us. What we ended up building was a purely fanciful dry-laid structure, more in the realm of an imaginative "folly" than a serious historic reproduction.

The stagecoach house is actually a pretend ruin. It was never a building and it never had a roof. The

place had never been fitted with windows and doors. No stagecoach ever stopped there. The parts of the structure that appear to be missing had not somehow disappeared or fallen down over time. The new, old-looking stagecoach house was merely a kind of set, like in a movie.

The attraction of building faux ruins has a lot to do with theater. It appeals to that curious tendency some of us have to pretend. What we constructed off in the forest was make-believe history. We were totally into the idea that the stage had to be set. The

**A large triangular-shaped stone presented itself as the perfect corner shelf for our stagecoach house. This sort of thing happens all the time: The material dictates the design.**

# Sarah's Castle

For some time now, I have taught a workshop on stone walls in Rochester, New York. During each visit, the central student project — a ruined castle — is enlarged. There are now arches, walls, benches and a turret with cantilevered stone steps leading up to a raised lookout area with a half-Gothic window. As the castle slowly grows, it is testament to the fact that, given a little encouragement and focused training, ordinary people can learn to make some truly inspiring structures with dry stone. All the project stones have been found and gathered locally from the Rochester area, and very little shaping is done to them. Our method is to find the best shape to fit the space rather than to try to change the shape of these field stones. Occasionally, we have to hammer apart large stones to create sharp, broken hearting stones — the small round stones gathered in the fields are not suitable for wedging the builder stones effectively, nor would they provide adequate friction.

structurally, but what they portray is not intended to be a manifestation of masonry perfection. Whoever stumbles upon this stone enclosure hidden in the woods along the California coast experiences the drama of an event unfolding. Visitors catching a glimpse of a fake ruin somewhere off in the distance feel the need to interpret what actually happened on this plot of land in the past. The visitor steps into an unwritten story, into another time, into a building that looks old and familiar but is actually unfinished and waiting to be completed.

The stones appear to be suspended in time. Dry-laid stone looks far more alive than any other structural material. This sense of everything being in a state of change is felt much more deeply when we enter some half-fallen-down structure, as opposed to how we feel when we come across finished stonework that is in good repair.

Follies are perceived to be at different stages of decay. If they are done properly, a delightful sense of the passing of time is achieved. Perhaps there is even some uneasiness or sadness associated with the mystery surrounding these faux structures. The stones, if laid right, in keeping with the surroundings or the theme of the landscape, produce a backdrop for our emotions. Seeing evidence of some mysterious edifice of stone off in the distance, we become aware of mixed feelings of awe and curiosity. As we approach and are surrounded by the stones, we might experience pangs of loneliness or longing for the past, even though the stones are merely playing the part of being old. Stones are good at evoking a feeling of nostalgia.

**Opposite: A recently completed dry stone folly.**

walls slowly took shape, in two three-week phases over two winters, and plenty of imagination went into the creation. In many ways, though, the building designed itself. It was more like an improvisation than a planned construction. The spontaneous quality of the piece would ideally give those who came to see it a reason to pause and reflect on the contrast of the temporary nature of a building, juxtaposed against the permanence of the stone material itself.

In a garden folly like this one, the stones are all given supporting roles. They are there to ignite the imaginations of those who come upon the stagecoach house. The stones are all fitted together

**Let's replace the kind of "ruin" pictured at right with a more appealing structure, such as the one that appears on the previous page.**

In this capacity, the act of building dry stone walls is a kind of performance art. Each stage of the ruin is a theatrical process. The wall builder is called upon to create a setting. The scenery is permanent, but the structure is temporary. It is part of an ongoing play that will be held over for a long time. The stones are called upon not just to stay up but to look like they have fallen down too.

Shakespeare wrote in his pastoral comedy *As You Like It* that, "All the world's a stage, and all the men and women merely players; they have their exits and their entrances." Whether it be a Gothic arch entrance or post and lintel, the characters entering the world of dry stone ruins get to imagine their parts in different ways and in different times. The mood of the place affects our emotions and makes us act differently. Like a piece of music written in an old style, perhaps a Celtic ballad, the well-crafted old-looking yet recently built ruin can legitimately create a sense of sadness and sweet longing.

Stones make themselves available to us in phases. As children, we may be attracted to them on a very simple level: We skip them, tumble them and bring them home in our pockets. Then the material takes on more meaning, until over time, we see its various uses in buildings, bridges and walls. We recognize stones have inherent strength, durability and resistance to aging. When we walk into a place where stones have been imaginatively assembled to give the impression of time, we become the players in the story they tell.

## A Place for Ruins

THE TWO PHOTOS on the opposite page show two very different masonry structures that visitors to our town might see if they strolled around for a while. One is a concrete stone building along the river that must have once been a utility building of some sort. The other is a dry stone folly built by wallers from Scotland, Switzerland, the United States and Canada during our second Rocktoberfest back in 2005.

Both are relatively modern ruins. Both structures catch your eye. Both create a sense of place and a feeling of time. Neither serves any particular use ... or do they?

Canada and the United States, relatively young countries in the New World, understandably do not have nearly as many ruins as countries in Europe. Old buildings are what give a place character. So we need to preserve the ones that we do have and, if we can, tastefully create new ones.

**Author Farley Mowat supervises the building of a "new" old ruin.**

## Fragmentation

CANADIAN WRITER, ENVIRONMENTALIST and activist Farley Mowat practiced a tradition in which "fiction" and "truth" were not mutually exclusive. We should remember and learn from stories told around the kitchen table or cooking fire, but they are not necessarily historically accurate in every detail. Mark Twain said, "Never let the truth get in the way of a good story." Mowat's maxim was similar: "Never let the facts stand in the way of the truth."

In a way, there's a parallel here with the construction of make-believe dry stone ruins: I try not to let the emptiness of today's landscape get in the way of

# Alban Beacon

In August 2006, we built an Alban beacon on the Canadian Atlantic coast. The installation was erected for Farley and Claire Mowat on their summer property near St. Peter's, Cape Breton Island. We gathered suitable rocks from below the high-tide mark and assembled them in a unique 8-foot-tall dry stone pillar on a point of land that could be seen from their house.

The beacon was made to replicate one of the many dry-laid structures built by a pre-Viking people who, according to Farley Mowat, came to Canada in fragile hide-covered double-enders from early Britain in search of walrus skins and tusks.

According to his fascinating book *The Farfarers*, many ancient conical piles of stacked stones, or "Tower Beacons," can still be found in Greenland, Labrador, Newfoundland and Arctic Canada. They stand from 7 to 14 feet high and range from 4 to 6 wide, and each has a unique shape. Often, these towers are found in groups or pairs, and they are believed to have been erected as markers. The beacons were different enough in shape and in style to help ancient seafaring people find their way around the coast.

This new dry stone beacon not only adds a sense of history and mystery to the rugged Canadian coastal landscape but stands as an accessible tribute to our prehistoric dry stone heritage.

The computer age leaves no ruins; the data is either there or it's lost, only to be found by computer technicians. The fragmentation of files and hard drives may not be something we can relate to, but stone ruins can provide that middle ground where we come to terms with the gradual fragmentation of this fleeting world. The past is always gone, and yet there is value in contemplating the loss.

We may not know the meaning or what it is we are remembering, but we absolutely need reminders of the past.

## Well-Built Obsolescence

PART OF THE allure of dry stone ruins is their rustic beauty. Many seekers of such an aesthetic simply like being around stone follies and ancient-looking garden structures, mostly because the stone they are made of is such an attractive material. Stone ages well, as opposed to the less-attractive way in which other materials quickly start to decay and deteriorate.

Generally, however, people don't like too much dilapidation. Broken plastic, wet cardboard, warped plywood and rotting beams are not beautiful or inviting. Stone alone has a dignity in decay.

It is interesting that many masons, even dry stone wall builders, are reluctant to produce something that looks less than new and perfect. It is as though they are not at peace with the fact that, inevitably, even stone is not indestructible. Stone is not forever. A dry stone wall builder who is able to make a structure

**This dry stone "cone tree" in Hood River, Oregon, was built on a round concrete base with a center pole that provided a fixed axis to scribe the ever-decreasing arc of the cone.**

creating a structure that suggests something meaningful had gone on here in the past.

Author Joel Knight wrote, "There is, I find, something very evocative about ruins — particularly recent ones."

genuinely look like a ruin is not insecure. He is not afraid of exploring the effects of time and dedicating some of his attention to celebrating the limitations time imposes on the most durable of materials. There is a satisfaction that comes with building something that looks like it has come to terms with mortality. No matter how well we masons build something, we know it will eventually collapse.

The thrill of the hunt is something we enjoy and cherish; any opportunity to search out and explore old ruins. We are looking for that hint of history, even in stone buildings that are not so old. This is partly because the stones are so old already that they have a history of their own. Combine that quality of stone with a patron who has a vision for what is appropriate on his property and a mason who knows how to cleverly fit the stone puzzle together so that it enhances the feeling of age, and you have a very believable, very attractive tribute to the ephemeral quality of life. There is the contrasting marriage of durability and impermanence.

The craft of aging something, whether it is a painting or a piece of wooden furniture, to make it appear older than it is, is a very demanding one. Rev. William Gilpin, in his essays on the picturesque, said the following:

"There is great art, and difficulty also in executing a building of this kind. It is not every man, who can build a house, that can execute a ruin. To give the stone it's [moldering] appearance — to make the widening chink run naturally through all the joints — to mutilate the ornaments — to peel the facing from the internal structure — to [show] how correspondent parts have once united; [though] now the chasm runs wide between them — and to scatter heaps of ruin around with negligence and ease; are great efforts of art; much too delicate for the hand of a common workman; and what we very rarely see performed."

To accidentally come across a well-built stone ruin is a magical experience. To be given the opportunity, and to have the ability, to make such a relic is a special privilege. Those of us who work with stone have the responsibility to do it right. If it looks too new or too perfect, it will defeat the very purpose of a folly.

# Dry-Laid Stone Hut

The design for this dry stone hut is based on buildings with vaulted roofs found in parts of southern France. Our circular hut measures 16 feet across, and the walls are nearly 4 feet wide at the base, leaving an inside chamber that is 8 feet in diameter. Large, flat stones were lifted up on to the roof and set in place in diminishing circles over the top of preciously laid corbel stones. The stones were not supported in any way from below as the dome was being built. This is an entirely dry-aid structure in which no mortar or wood is used. The roof sheds water and keeps the inside perfectly dry. It is now used as a garden hut and makes a very cozy shelter from the rain and wind.

# The Frontenac Arch

This dry stone arch was built at the Frontenac Arch Biosphere Office near Gananoque, Ontario. It was the first permanent dry stone arch to be built on public property in Canada. The students who took part in this project all had a chance to try their hand at shaping the rugged, chunky sandstone material to create the voussoirs and quoins (the term for masonry corners). This unique structure was completed in just two days using 16 tons of random stone.

# Chapter 5

# Life Lessons

I WAS INTERESTED to discover that Carl Jung, the leading pioneer of psychology and someone steeped in the study of the mind and human behavior, maintained a continuing hands-on interest in the mysterious connection between stones and human creativity. This is what journalist, author and BBC broadcaster Mark Lawson says about Jung:

> Carl Jung had a drive to transmute psychic images into form and substance. It was the force behind Jung's tower-building impulse in Bollingen. Through it, he participated in an ancient and archetypal urge to secure the ineffable in the permanence of stone. History is filled with the archaeology of such longing, from the Paleolithic Venus of Willendorf, the stone circles of Northern Ireland, Stonehenge, and the monumental heads of Easter Island. At the center of Islam is the black stone, set in the wall of the Ka'ba at Mecca. All holy places and shrines share this human urge to meld meaning into rock so that it may endure for generations.

Throughout his life, whenever he got stuck, Jung "hewed stone." Stonemasonry was his therapy. After his wife's death in 1955, he wrote that "the close of her life, the end, wrenched me violently out of myself. It cost me a great deal to regain my footing, and contact with stone helped me."

## The Creative Act

BEING CREATIVE IS the highest form of human fulfillment. It beats any other kind of "re"creation. It is the opposite of so much of what we do, that is, wrecking creation. Some might argue that it is more important than procreation.

The creative act is vital. No amount of shopping, education, entertainment, competitiveness or commitment to selflessness or selfishness will secure the sense of pleasure and purpose that comes with being involved, no matter how briefly, in a creative act.

We are all meant to be creative. It is part of who we are. There are many ways to be creative, but stone provides the rawest potential for human beings to explore their creativity.

Stone exists in abundance as the silent, unformed, motionless protagonist able to unleash that inner artist in many of us. There is a bond between us and stones. We have more of an affinity with them than we know.

When we come upon a stone, we must learn to see it as more than a blank canvas. The very mass of the stone contains within it an endless source of possibilities on which we can build.

Being creative isn't something you can just pick up, especially if you are not currently using your hands. Perhaps stonework can be a kind of transition on the way to our becoming more creative. As we place and fit stones in a wall, our mind discovers solutions that it didn't know were there before. In moving stone, we are learning how to express ourselves, and this is the beginning of art.

When facing a problem, remember that while there are no perfect answers, there are enough good solutions to be able to make a good wall.

It is important to embrace this concept, instead of being intimidated by all the wrong choices. Don't let your mind tell you that it will never be good enough. Be willing to make mistakes, as I am, and encourage everyone to risk failure. There is always someone out there who is going to throw stones at your work, instead of helping you build with them. Creativity is a reward in itself, and the feeling of accomplishment is well worth the effort, even if people (and even you) are in a hurry to question your credibility and your skill level.

## Letting Go

AS HARD AS it is to learn skills and acquire useful knowledge, letting go of things is even harder. We are not good at letting go of bitterness and resentment. Our minds have too much of a sense of being wronged. We hold things in mentally and emotionally, and they eat away at the very framework of who we are.

On the other hand, things we hold on to physically are much easier to let go of when it's necessary. A hot cooking utensil, perhaps. A sharp object. A rock that isn't useful or even a stone that is. No doubt we can learn how to let go of the things in life, things that are weighing us down or to which we cling, by watching how our hands let go of things and even drop them. Our hands are good at it.

The heavy awkward stones are only in transition when we pick them up to build a wall with them. It becomes far too tiring otherwise. Our hands are being used, not just to pick up these stones and move them, but to release them as well. Nor is it our job to forever hold them in place in the wall or to keep coming back to reposition them. Good structure depends on our maximizing the friction and center of gravity of each stone and making sure the stones are nestled securely into each other within the wall. We are not just trying to stack them in some temporary balancing act and hover round them with our hands, wondering when they're all going to fall over. If we are good at walling, we are good at putting things where they need to be, and at letting go.

The letting go part is not abstract or gradual. It is definite and precise. And more importantly, it is with an understanding and a complete appreciation for function. Our minds might understand "purpose," and we may resolve to be a better person, but it is our hands that demonstrate the higher level of commitment, which is function. We are making a wall, not theorizing or imagining. If we think with our hands, we "think" with something different than purpose. Even if we make mistakes, it is not on purpose, but it is part or how we function. If people criticize us for what we do, and we have a tendency to hang on to our hurt, we can still function by finding the right place for that awkward, seemingly useless stone and putting it in its place. It will become an integral part of that very thing we are building, a beautiful wall that is made up of many, many similarly awkward stones that, once they are in position, merge into the wall and then are individually forgotten. The hands are very good at forgetting. It is the whole hands-on, hands-off process that can produce not just a wall, but a work of art.

MUCH OF WHAT I believe about the benefits of working with stone is expressed in the following poem.

What creates a sense of place?
A glade, some shade, a quiet space?
Enclosed protected but not confined
Unthreatening walls yet well defined

A gateway there through which we pass
To contemplate the things that last
Surrounding walls affirming love
Separate from and yet part of

The nameless regions of our wide expanse
These walls define our circumstance
We reminisce we soar we sigh
Between the womb and where we die

A tranquil haven lined with life
Edged with meaning, free from strife
The landmark milestone resting spot
Where space embraces time until it's not

And we have paused to know we're "there"
A place unique from everywhere
A resting spot flanked by stone
A space within a sheltered zone
A memory held within a spatial song
Where we can feel that we belong

What creates a space like this
Where old men rest and lovers kiss
Where wasteland ends and purpose thrives
Where silence sings and peace survives

Such tracts of land, such sacred ground
Are very rarely ever found
And rarer still can they be made
Except by stone on stone securely laid

A wall whose stones are made to fit
As if they were always part of it
Lend strength and certainty to such
As we who are sensitive to spatial touch
We wait and grasp our new belonging
For sense of place is what we're longing

—*John Shaw-Rimmington*

The dry stone feature, opposite, was designed for the horticultural school in Niagara Falls, Ontario. I was asked to create something unique during the three days of the Niagara Garden Show using the stones they had available at the park. *Cheese Wedge* was going to be left standing after the show. Often we build things at demonstration event that are taken down afterward.

Our goal was to create something unusual, a piece that would generate interest in dry-stacked building in Canada that went beyond stone walls. The whole event was a great success, and *Cheese Wedge* remains the pride of the school.

Historically, a stone boat was a low-profile horse-drawn sled used by settlers to haul huge stones off to the sides of their freshly cleared farmland. In Port Hope, the stone boat has come to take on a different meaning altogether. In 2006, a structure was built to commemorate a book by the internationally acclaimed author and local resident Farley Mowat. His book, *The Farfarers*, about a pre-Viking people who came to Canada in search of walrus hides and tusks, was the inspiration for the dry stone boat. It replicated the shelter made by the Alban people who overwintered in parts of Arctic Canada by building dry stone enclosures and roofing them with their upturned double-ended ocean-going crafts. As the simulated walrus skins of the hull fade, this new stone boat has started to look even more authentic.

**B**uilt in 2008, this installation was designed to create the impression of a sunrise. The large granite boulder, an erratic, must have weighed at least two tons and was lifted into the wall with a large crane equipped with straps and chains. The arch, which straddles both sides of the boulder, was constructed to look as if it was self-supporting and not actually touching the "sun." One lone boulder, or even two or three, seem not to have any reference to their surroundings. Introducing smaller stones (especially in a formal pattern of a wall or arch) accentuates the size of the bigger stones and gives some order and perspective to the composition.

The mixed granite fire pit seating area, opposite, utilizes predominantly local round stone from a quarry not a mile away. The owners wanted a fire pit that wouldn't look unsightly when it wasn't being used, and they wanted to incorporate some of the local stone they had gathered themselves. The 10 acres behind the fire pit provided a beautiful natural backdrop. The stone seats were procured at a quarry where metamorphic rock comes out in beautiful, flat, slabbed layers. The stones in the burning pit were placed vertically in the sand in a radiating pattern. The heat from the fire glances diagonally off the stones rather than hitting them straight on as they would if they provided a perpendicular border to the flames.

The stones in this outdoor fireplace are dry-laid in a cladding around a mortared block and firebrick firebox built onto a concrete pad. The final structure has all the rustic charm of dry stacking yet maintains the functionality of a brick- or stone-and-cement structure. The arch is freestanding — there is no lintel. Three stacks of 4×4 clay flues are mortared in place, around which stones have been placed to create the chimney.

There are many different styles of dry stone walls in Britain. This wall in the Lake District has a unique type of decorative coping. Beck stones. or river rock, and local slate are alternated, not only in the layers of the wall itself but in the top copes. Perhaps there were not enough flat cap stones to do the top entirely of slate. Perhaps this was just a more interesting way to finish the walls.

**A** side view of *Rubble Helix*, two twisting columns of limestone held together with massive through stones.

**T**he dry stone *Venus Gate* amphitheater and plinth, opposite, were built near Caledon, Ontario, in October 2011 by professional wallers from Scotland, Wales, England and Canada. The plinth echoes the pointed oval opening to the amphitheater. In this way, a dry stone wall represents an ongoing conversation between positive and negative space.

**Previous spread:**

This wall is built of quarried dolomitic limestone. The early spring flowers in the foreground and the trees in the background seem to embrace the beauty of this new addition to the landscape.

A friend of mine from Virginia became interested in building dry stone walls on his property. He lives in an area where there are many beautiful old walls that have been made with local stone and which incorporate lots of interesting styles and features.

After an excavator had worked full time on his property for several months, my friend wrote to tell me he had finally uncovered and dislodged enough bedrock and loose stone material to do something interesting.

I kept receiving cryptic e-mail messages from him telling me how it was going.

They went something like this: "The wall is about 10 or 12 feet wide now in parts and is taking up a lot of material. It's a good thing I have lots of it." I checked to see if he didn't mean 10 or 12 feet long?

"Nope , in fact it must be nearly 30 feet wide now along one part, and I figure that I won't fill it in but sort of let it have a kind of rounded open area in the middle of the wall."

Until I saw it for myself, I couldn't imagine what he was talking about.

The original Bedford Street retaining wall had collapsed, and there was not enough stone to make the walls thick enough to resist the heave of the frost. We rebuilt it using twice as much material and created this beautiful, sturdy wall.

In regular horizontal-coursed walls, stones laid flat have only their own weight (and the weight of some stones above them) to limit any shifting. But stones laid diagonally, as in the herringbone pattern we used in this wall in Victoria, British Columbia, are further held in place by the weight of all the stones leaning to the left or right of them. Stones in a herringbone wall are more connected and their dependence on each other is structural.

An old stagecoach road runs through my client's property in northern California. In 2009, I was asked to design and build a stagecoach house there, as a kind of fanciful reminiscence of what might have been, in the style of a dry stone ruins.

This folly's design changed as the sizes and shapes of material we were able to bring up the stagecoach road began to dictate how the structure needed to be built.

The stagecoach ruins were completed in 2010 and have now been seen by many visitors; photographs of the ruins have also appeared in several books.

These ruins have taken on a special meaning for me. They represent my transition from traditional walling to envisioning and constructing imaginary garden installations and, ultimately, various structural sculptures.

There are those very rare times when you see the pattern in the abstract forms of nature. Things briefly lose their random appearance. An unmistakable design appears amid the usual random complexity and strange asymmetry of nature's handiwork. For me, looking at the herringbone wall, opposite, triggers the same kind of magical moment where I think, "There is a hidden logic to the design I'm seeing," a kind of secret pattern there, waiting to be discovered.

This dry-stacked Christmas tree near Rochester, New York, is made entirely from Medina sandstone. Rings of longer stones were counter-balanced in a circular pattern, stacked and spaced with blocky, shorter, stones between each layer.

**S**tonemasons have built an almost perfectly oval doorway into this dry-stacked stone wall.

**T**he low wall, opposite, runs along a private property in Vermont. Its irregular lines create additional strength in withstanding the pressure of a raised garden.

# The Kerry Landman Memorial Tree

EVERY NOW AND then, inspiration, talent, stone and the stars align in such a way that something magnificent happens. In the primal world of stone walling, where no mortar or glue or any kind of industrial advantage is tolerated, even the most well-built structures rarely climb out of the category of traditional craft to become (dare I say?) works of art.

If fitted skillfully, stones can be pretty much be relied upon to support and create an impressively solid and long-lasting form, but rarely do they get the opportunity to perform as they do in Eric Landman's tree installation in the Dodds & McNair Memorial Forest near Orangeville, Ontario. Dedicated to his wife Kerry and built primarily by Eric and the couple's son Jordon Mason, the wall speaks to the heart on many levels. Since its completion in March of 2012, one or two images of the work have exploded across the digital universe. I suspect that by now, millions of people have gazed at it, recognizing it to be something of a masterpiece. And yet like all masterpieces, it really needs to be seen in person in order for its full meaning and materiality to be appreciated.

Standing there in the snow, I see cold, solid, everyday stones transformed into a living mosaic. They combine to create an impressionistic sculpture, a natural synthesis of uplifting beauty. The image of the tree bursts through the inanimate restrictions of mass, gravity and geology.

The piece is bedded in the landscape. Placed in a gallery of nature and lit by beams of golden sunlight that cascade through the dappled forest backdrop, this tree appears to be growing. Indeed, the piece grew out of love "from the ground up" during the cold winter after Kerry's death.

Unlike a painting or a mosaic, in which any part of the whole image can be worked on at the same time, the outline of the tree could only be "grown" upward, stone upon stone. Care had to be taken to ensure that some of the stone foliage was built into the wall in places where it did not yet appear to be connected to the whole. This is the reward and the challenge of creating credible representations within the dry stone medium.

The overall effect is overwhelming; as a freestanding stone feature, this piece is an incredible accomplishment. Nearly 5 feet deep at the base and over 12 feet tall, it is a totally self-supporting structure, visitors can't help admire the surface tree image even as they reflect on the interconnectedness within—the inner beauty and depth of emotion that went into creating this enduring piece.

Previous spread: Author John-Shaw-Rimmington helped master mason Norman Haddow build this wall in Aberfeldy, Scotland.

The Kerry Landman Memorial Tree, opposite, is a spectacular wall that Eric Landman and Kerry Landman's son Jordan Mason built in her memory in 2012. Local rounded granite fieldstone were used to represent the leaves, and the moss-covered stones add to the effect.

# Acknowledgments

STONEWORK OF ALL types lends itself to rich collaboration, both physical and creative.

A number of the projects in this book were brought to life with the much-appreciated professional assistance of Sean Adcock, Norman Haddow and Patrick McAfee.

Thanks are due to many other people for their help and inspiration as well. They include Thea Alvin, Leigh Bamford, Christopher Barclay, Menno Braam, Patrick Callon, Dave Claman, Scott and Cindy Cluett, Andrew Currie, Kenny Davies, Sean Donnelly, The Ennismore Horticultural Society, Georges Foliot, Scott George, Katherine Gleason, Anne Halliday, Dylan Hansen, Mark Harris, the Hendersons, John Henry, Andrea Hurd, Akira Inman, Eric Landman, Tomas Lipps, Dean McLellan, Joe Mitchell, Claire Mowat and the late Farley Mowat, Brian Mulcahy, Peter Mullens, Christopher Overing, Evan Oxland, Mike Patten, Dan Pearl, John Redican, Matthew Ring, Gavin Rose, the folks at Sara's Garden, John Scott, Colin Shaw-Rimmington, Stephen Smith, Sean Smyth, Dan Snow, Trevor Spik, Amanda Stinson, John Thompson, Danny W and Dylan Wales.

Particular thanks are owed to Mark Ricard.

And a big, big thank you to my wife, Mary.

Ultimately, the real heroes of many of these projects are the private clients who wanted to have long-lasting artwork built of solid stone. To these unnamed people, let me say that I am very grateful to have had the opportunity to help you envision and realize your dreams.

# Further Reading

Brooks, Alan, and Sean Adcock. *Dry Stone Walling: A Practical Handbook*. British Trust for Conservation Volunteers, 1999 (ISBN: 0-9467-5219-2).

Donovan, Molly, and Tina Fiske, John Beardsley and Martin Kemp. *The Andy Goldsworthy Project*. London: Thames & Hudson/ National Gallery of Art, 2010 (ISBN: 978-0-500-23871-4).

French, Lew, and Alison Shaw (Photographer). *Stone by Design: The Artistry of Lew French*. Layton, Utah: Gibbs Smith, 2005 (ISBN: 1-58685-443-7).

Gardner, Kevin. *The Granite Kiss: Traditions and Techniques of Building New England Stone Walls*. Woodstock, Vermont: The Countryman Press, 2001 (ISBN: 0-88150-506-4).

Goldsworthy, Andy. *Wall at Storm King*. New York: Harry M. Abrams, 2000 (ISBN: 0-8109-4559-2).

———. *Stone*. New York: Harry M. Abrams, 1994 (ISBN: 0-8109-3874-2).

———. *Arch*. New York: Harry H. Abrams, 1999 (ISBN: 0-81009-1993-1).

Laheen, Mary. *Drystone Walls of the Aran Islands: Exploring the Cultural Landscape*. Cork, Ireland: The Collins Press, 2010 (ISBN: 978-1-84889-012-1).

McAfee, Patrick. *Irish Stone Walls, History, Building, Conservation*. Dublin: The O'Brien Press, 1997 (ISBN: 0-86278-478-6).

McRaven, Charles. *Stone Primer*. North Adams, MA: Storey Publishing, 2007 (ISBN 978-1-58017-669-9).

Rainsford-Hannay, Colonel F., *Dry Stone Walling*. London: Faber & Faber, 1957.

Reed, David. *The Art & Craft of Stonework: Dry-Stacking, Mortaring, Paving, Carving, Gardenscaping*. New York: Lark Books, 2003 (ISBN: 1-57990-218-9).

Scully, Sean. *Walls of Aran*. New York: Thames & Hudson, 2007 (ISBN: 978-0-500-54339-9).

Snow, Dan. *In The Company of Stone: The Art of the Stone Wall*. New York: Workman Publishing Company, 2001 (ISBN: 1-57965-347-2).

———. *Listening to Stone*. New York: Workman Publishing Company, 2008 (ISBN: 978-1-57965-371-2).

The Dry Stone Walling Association of Great Britain. *Dry Stone Walling Techniques & Traditions*. Cumbria, U.K., 2004, 2008 (ISBN: 0-9512-3068-9).

Thorson, Robert M. *Exploring Stone Walls: A Field Guide to New England's Stone Walls*. New York: Walker & Company, 2005 (ISBN: 0-8027-7708-2).

———. *Stone by Stone*. New York: Walker & Company, 2002 (ISBN: 0-8027-7687-6).

Author John Shaw-Rimmington has two websites. One is a daily blog titled Thinking With My Hands, in which he reports on things having to do with building. You can find it at **thinking-stoneman.blogspot.ca**. The other is called Dry Stone Walling Across Canada (DSWAC) and can be found at **dswac.ca**. This is a regularly updated site that lists events and workshops taught by the author and other instructors in Canada. It also showcases Canadian walling content and features both new and historic work, as well as occasional articles of interest to DSWAC members.

For more information, please contact:
Dry Stone Walling Across Canada
4 Cumberland St.
Port Hope, Ontario
Canada, L1A 1Z5
Email: mcclaryharris@sympatico.ca
Website: www.dswa.ca

# Index